# RISK & SAFETY – What Really Counts

Edition 01/2017

ISBN 978-3-9524178-5-0

*Those who would give up essential Liberty,*
*to purchase a little temporary Safety,*
*deserve neither Liberty nor Safety*

Benjamin Franklin, U.S. writer, philosopher, politician, 1706 – 1790

# Contents

# 1  Preface

Many years ago, I spent a few days in the mountains, enjoying outdoor life and ski touring. After previous snowfalls, the avalanche hazard had been assessed as possibly critical on that particular day. Nevertheless, I decided to explore the terrain, leaving my friends behind. While ascending with my skis, suddenly a slab broke loose just above me. My heart beat rate doubled. The whole slope started to move under my feet. I found myself sliding without any control, my skis dragging me down like anchors, the heavy snow overwhelming me up to my waist. Since I was able to write this book about a quarter of a century later, you might presume that the outcome of that little adventure, which is described more in detail in chapter 8.1, was fortunate for me. In that time, I considered myself an experienced mountaineer, knowing about the basic safety measures to avoid avalanches, but I disregarded some of them deliberately. Do you know about the thrilling sensation of skiing down a steep slope in pristine powder? Probably you will understand the fascination of risk. It's about emotions, not about rationality. Young people in search of adventure are exposed to extraordinary risks.

Later I became involved in risk and safety issues on a professional level. Tunnel safety and ventilation has been my primary field of profession in the past two decades. This book arose from a technical article in 2011 with the ominous title 'Reducing Costs and Improving Safety of Road Tunnels' [108]. To that, I got positive reactions from many professionals, although my recommendations were obviously neglected in practice. A public servant, being in charge of investments in road safety in a 'not to be named country' didn't mince his words, explaining that reducing costs and improving safety might be officially communicated goals towards the public, but honestly do not provide real incentives neither for procurers, nor for suppliers. Leaders, government officials and company managers applying thorough analysis of benefits, costs and risks for thought-out decisions may exist in Disneyland, but not in the real world.

The deeper I got into the matter, the more I was astonished about the gap between public risk perception, officially communicated policy, and the mess in the harsh reality. Working under time and cost pressure, failures and accidents that could be easily avoided happen regularly due to design flaws, bad workmanship, erroneous operation, and mostly lack of funds, time and will for proper analysis, evaluation of good solutions and practical quality assurance by thorough testing and improving. I describe many case studies, trying to write in an understandable way to the general public, without going into the technical details.

Have you ever wondered why air quality in many road tunnels is quite poor, and why fires in tunnels often lead to the whole tunnel getting filled with smoke, endangering and sometimes even killing people, despite expensive tunnel ventilation equipment? In contrast, how many casualties do fires in tunnels really cause? What would you do in such a situation?

Fig. 1    In a tunnel filled with smoke

The worst European road tunnel catastrophes, namely the Mont Blanc tunnel fire disaster with 39 fatalities, the Gotthard road tunnel fire with 11 fatalities, and the bus crash in the Sierre tunnel with 28 fatalities (among those 26 children), were caused by Belgian trucks and busses. Obviously, the most effective way to improve road tunnel safety would be to ban Belgian vehicles from Alpine roads. Or is that a premature conclusion, based on hindsight interpretation of a random correlation? Are some of the new design standards that have been invented as consequences of the mentioned tunnel disasters, leading to investment of billions of Euros of taxpayers' money in tunnel safety, based on premature conclusions too?

Even when most people in modern affluent societies live as safe as never before in human history, there is still potential for improvement, and we must avoid to relapse into medieval conditions due to ignorance, carelessness and ruthlessness of particular stakeholders. It is only human that decisions with far reaching consequences are mostly based on randomly chosen particular arguments and simplified cause-effect templates. Proven engineering principles to manage technological risks, as described in chapter 12, seem to be neglected in favor of bureaucratic empty runs, following dubious standards and codes without questioning them. From an overall point of view, safety is impaired when time and funds are wasted for safety measures with limited usefulness, and subsequently those funds lack when more important tasks are at stakes.

# 2 Basics

## 2.1 Münchenstein Railway Disaster

Münchenstein, south of the city of Basel in Switzerland, had been the scene of the worst disaster in Swiss railway history[1]. The 42 m long bridge crossing the Birs river had been originally designed and built in 1875 by Gustave Eiffel, who later became famous for building the tower in Paris that still bears his name[2]. On June 14, 1891, an overloaded train with two locomotives was passing over the bridge, which suddenly collapsed under its weight. The front part with two locomotives, four passenger wagons and two freight wagons fell into the raging river. 73 people died and more than 150 were injured, even though in the rear part of the train, the passengers got away without serious damage.

Fig. 2    Collapsed railway bridge in Münchenstein *(Public domain)*

The investigating commission found various possible causes for the failure of the steel bridge. An important issue was that the bridge obviously had been underdesigned, without taking into account sufficient safety margins for its stability. The quality of the steel was not in accordance with the required speci-fications. Further, the bridge had not been adequately repaired and reinforced following damage by a previous floodwater in 1890.

---

[1] Münchenstein had also been the domicile of tennis legend Roger Federer, and other, not so famous citizens, for instance the author.
[2] Unlike the bridge, the tower did not collapse so far.

Those issues were obvious consequences of financial pressure and profit maximization by the privately owned railway company. In the years after the disaster, all Swiss railway bridges were examined, and some had to be reinforced. The private railway companies were put under stricter government control, and in the following years, most of them were nationalized. Most importantly, national standards for the design and layout of bridges, prescribing defined safety margins, were elaborated and implemented. In those days of development of mass transport facilities, such standards were important in establishing an adequate level of transportation safety. Today, most bridges in the developed world are way oversized, and rarely ever collapse. Nevertheless, previously unknown effects may still cause failure, for instance the collapse of the Tacoma Narrows Bridge in the U.S. in 1940 due to aeroelastic flutter.

Fig. 3    Collapse of Tacoma Narrows Bridge *(Public domain)*

Bridges are static structures, developed during thousands of years. In many fields of application, the situation is different, as will be described in the following chapters. Modern technologies, industrial and power plants, means of transport, and related safety equipment are a rather complex issue, and safer – at first sight – is not always better.

## 2.2   Safety and Risk

*'Everyone has the right to life, liberty and security[3] of person'*

(United Nations, The Universal Declaration of Human Rights, Article 3)

When being asked the question 'Am I safe?', most people answer to a different question: 'Do I feel safe?' To answer the first question, a profound analysis would be necessary. In colloquial speech, risk is feared as a hazard, leading to possible damage. According to the ISO vocabulary [59] and successive norms [59], [61], risk is the effect of uncertainty on objectives, an effect being a deviation from the expected - positive or negative. Therefore 'negative risk' leads to possible damages and 'positive risk' to possible chances. However in this book, the term risk only refers to hazard with possible negative impact, as is usual in common understanding.

Feeling safe is what most people strive for, but while feeling safe, avoiding obvious, but possibly acceptable risks, they may be vulnerable to much higher, but hidden risks. Risk awareness is a compromise between carelessness and paranoia. Risk analysis may be a useful approach to evaluate an appropriate, previously defined level of safety, but there are some fundamental limitations. We try to imagine what might happen, defining possible scenarios and assigning corresponding likelihoods. Some scenarios might be unique, some may never happen, and some are unknown in principle. Those that nobody has thought about before might lead to the worst disasters.

Risk is not only limited to immediate danger of life and limb, but to health and wellbeing in general. There are risks to the environment, which in turn threaten our health and life and those of our descendants, financial risks in any project or investment, and many risks where the possible damage is not so obvious at first sight, for instance the decrease of biodiversity.

---

[3] Safety and security are basically synonyms. The two words differ in connotation and are often used in a different context. Safety is the state of being safe, that is without unacceptable risks, which would be achievable in a defined context, for a limited time, with certain specific assumptions. Security is usually comprehended as the means and ways of preventing, detecting, and responding against deliberate damage inflicted by people, like unauthorized access, sabotage, theft and seizure.

## 2.3    Benefits, Costs and Risks

In any activity, humans usually pursue benefits. Unfortunately, any issue with a benefit is accompanied by unwanted side effects, costs and risks. Decision making is about weighing whether the benefit is worth the costs and the risk. Risk mitigation measures on one side increase the costs, but may (hopefully) reduce the monetized damage from quantifiable risks. The essence of any risk assessment are the following questions:

- What are the benefits, costs and risks?
- Is a risk worth the benefit?
- Are risk mitigation measures worth the resulting risk reduction?

The higher the safety, the more important the residual risk becomes. Many risks are accepted because the benefits are considered being worth it. Avoiding risks means avoiding chances. Which one is emphasized, is a question of point of view. In some risks, such a benefit is not always obvious, but abstract, for instance mere convenience, or 'adventure'. Sometimes humans take hazards simply because doing so 'feels good' at the moment, like eating sweets, smoking or driving fast and other thrilling activities.

Safety is not available for free. Mitigating risks is always a tradeoff against the primary purpose of any endeavor or facility. Such a primary purpose may be the accomplishment of a performance task, production of goods or energy for an industrial plant, providing means of communication for roads and rail, or earning profit for any company. Unfortunately, most measurements for risk mitigation require additional costs and time, and may hinder the primary purpose. Therefore, safety is directly related to wealth and abundance of appropriate means. A high safety level can only be achieved in wealthy societies. Even with many potentially dangerous technologies, for instance automobiles and industrial plants, the accident rate is quite low in affluent societies. In contrast, in poor, developing countries, such new technologies regularly lead to disasters.

Even when wealth is the basis for safety, losing money does not directly endanger your health or life; however it may lead to some nervous tension. Therefore this book is not about financial risks. Plenty of literature from competent authors about that issue is available.

## 2.4    Individual and Collective Risk

There is an important distinction between the individual risk from the point of view of a single person and the collective risk for a defined community. Individual risk can be expressed as probability at which a particular individual may be expected to sustain a given level of harm from the exposure to specific hazards. Collective risks are related to a particular group of people, technology or to a site, for instance a power plant or a road.

Collective risk implies individual risk to a number of exposed people. For the collective risk, the key parameter might be for instance the number of casualties, costs of treating injured/disabled people, species extinction, or financial losses for a given time or local frame. However, such numbers are to be understood as a mere estimation, since not all potential causes of death can be allocated to the consequences of the particular risk.

Risk mitigation measures don't always have the same effect on individual and on collective risk. In fact, mitigation of collective risk is often achieved by increasing individual risk to particular members of the collective, when some individuals are sacrificed for the wellbeing of their society. There are many issues where interests of individuals are contradictory to the collective interest.

An example is public health care in most Western countries. Statistically, approximately 80% of healthcare costs for stationary treatment are spent in the last 12 months before death. Since life becomes safer, and people generally become healthier and older[4], healthcare costs are shifted towards those older people. How could the increasing healthcare costs be limited? Shall people exceeding a certain age be deprived of life-extending measures? Could the money be spent in a better way to avoid suffering and pain to younger patients, who maintain the healthcare system by paying insurance premiums? Such questions might challenge our ethical standards

---

[4] Referring to West European countries – in the USA or in Russia, the average life expectancy is actually decreasing.

## 2.5   Human Behavior

Human behavior affects the probability of an incident to occur and also the extent of damage after the incident has occurred. Humans may cause incidents and as well increase damage by inappropriate reaction. And humans do harm to other humans, unconsciously or deliberately.

Understanding why people make mistakes and errors is essential for any risk assessment. This issue has been systematically investigated for instance by Reason [113]. One of Reason's conclusions is that latent failures are inherent in human organizations, resulting from inappropriate systems, design flaws, and hidden issues, and such impose usually a higher risk than 'active' failures, caused by improper operation, accidental misuse or sabotage. In a technical system, described by physical laws, the risk of failure can be theoretically calculated when probabilities of failures of each component are determined. Statistical laws apply. By introducing human behavior into the system, an element of principal uncertainty is added. Human behavior is irrational, especially in stress situations, and often unforeseeable. The influence of human behavior can be estimated only over a statistically significant number of people or over a long time.

As a tunnel safety expert, I have seen many records of fire incidents in tunnels from surveillance cameras. Fast smoke spread is a deadly threat to tunnel users, and would require quick escape from the danger zone. Nevertheless, many drivers prefer to stay in their cars. Reluctance to leave their propriety may eventually kill them. When a fire breaks out, people often stand around, watching the spectacle, and even filming with their smartphones. You see drivers deliberately leaving the safe zone and walking into the smoke. However, I know from my own experience that it is impossible to maintain full situational awareness and clear thinking under stress in dangerous circumstances. Some examples are described in this book.

Humans usually don't cause damage deliberately. Most people want to do a good job and survive any activity unharmed, even when risks are deliberately accepted. Accidents are often caused by people who make wrong decisions. But having been wrong can often be only determined in hindsight. At the moment of the decision, it was considered to be right. The decision maker had to achieve a goal for which he took risks that obviously were acceptable from his point view, taking into account the information that was available to him at the moment. Risks with consequences in the future are a delicate matter, even when people are fully aware of the hazards. As a general pattern, we prefer immediate reward, taking into account negative consequences in the future.

Weakness of will lets people smoke cigarettes even when knowing that it may cause lung cancer on the long term. On the other hand, people are forced to take risks by financial and time pressure, or by their superiors in hierarchic organizations.

Beside the fact that humans cause risks, their behavior is also essential for the resulting consequences. People suffer and die because of doing the wrong thing and behaving the wrong way. To minimize human error, standard procedures have been invented. Many technical systems are operated fully automatically, achieving a high level of safety and reliability. But automation does not eliminate risk completely; it rather shifts it to other threatening scenarios.

## 2.6   Media

The role of the media, like newspapers, TV and the internet is ambivalent. The Media provide an almost exhaustless source of information, and are important for the perception of risks beyond immediate experience. On the other hand, acquiring information exclusively from the media affects the ability of independent cognition and reflection. Moreover, modern media provide an overload of data and contradictory information, which can hardly ever be seriously analyzed. The trustworthiness of public sources is always questionable. What really happened, what has been concluded from the investigations, and what has been published are three different things, and you will never know the truth unless you have witnessed the incident personally – in fact, not even then, taking into account limited perception and subjective cognition.

Due to the representativeness heuristic, people fear what comes easily to their mind, which is mostly either from recent personal experience, from their social environment, or by the media. Journalists earn their living by describing real or imaginary hazards in a flowery language, emphasizing spectacle and sensation, and mostly focusing on exotic risks with high consequences, but low probabilities. Common risks with higher probabilities, affecting more people, do not get much public attention, for instance health risks like diabetes.

An incident may attract media attention and lead to high risk awareness in public opinion, even when the objective risk would be negligible.  As time passes, emotions die down, and carelessness increases. Previously taken risk mitigation measures are questioned and funds are cut subsequently. Then another incident happens and the hype starts again. Public risk perception continuously changes according to such a cycle process.

# 3 Hazards

## 3.1 The Measure of Damage

Any damage, whether referring to human, material or financial losses, consequently leads to unwanted suffering, injury or death of single individuals. Even damage to the environment or to other animals finally affects humans. Most people would define death as the worst case of loss. Does that make sense, since death is certain anyway? Death as a measure for consequences in human risk issues is defined more precisely as premature death due to an avoidable cause. However, many fates may be considered worse than death. Wouldn't you consider the death of your children or other beloved ones worse than your own death?

Biologically, life is about reproduction and survival, not about living as long as possible or individual happiness. From a practical point of view, you should be healthy, not suffer, and enjoy life's pleasures – and provide the same to your children. Philosophically, and most important for any higher developed life forms, life should have a meaning. But that would definitely exceed the topic of this book, since the meaning of life has been reflected and discussed for thousands of years. There are many different views, influenced by society, culture and religion[5]. Therefore, the 'worst case' for an individual would be to die after having lived a senseless life, full of suffering, without pleasure, and without leaving any descendants. But this doesn't necessarily mean a premature death. Which safety measures really aim at preventing such a 'worst case'?

However, effects like suffering or unhappiness are subjective values. In contrast, the death of one particular individual, or the number of casualties of an incident, provide a statistical quantity that can be measured, calculated and assessed in a simple way. The number of deaths is not the only figure that represents the impact of catastrophes, since many survivors may have lost their entire basis of existence. Taking into account such numbers gives us an idea about the extent of loss of the worst disasters. Other factors like injuries, material and environmental damage, resulting in the decrease of wealth and prosperity, should be taken into account as well, but are difficult to quantify.

---

[5] Religions introduced the idea of a 'life after death', by that also extending the possible risk scenarios. Promising paradise after death in exchange for sacrifice in the previous life is a powerful instrument of manipulation.

For instance the nuclear disaster of Fukushima did not lead to any immediate fatalities, but left an enormous radioactive contamination in the surroundings and the sea. These consequences can be evaluated by assumptions on future casualties, in particular thyroid cancer among affected people, and the financial losses due to contaminated areas, leading to the evacuation of people and outage of agricultural products and seafood. Such estimates can only provide rough numbers within a broad range of uncertainty.

Finally, in our society money is the ultimate measure, not only for material damage. Repair and replacement of destroyed infrastructure and equipment, deployment costs for emergency and rescue services, cleanup works, lost revenues due to disabled production sites, all this leads to directly calculable costs of particular incidents. In many lawsuits, abstract suffering is reimbursed by monetary compensation. Injury can be monetized by costs of treatment and missed income, but how can you express the pain in money? The answer might be given to a certain extent by psychological experiments in which the test subjects were reimbursed in relation to deliberate exposure to pain. It depends strongly on the circumstances. On the other hand, simple material damage can result in traumatic suffering, when you lose all your belongings and basis of existence. Human wellbeing can be closely linked to material possessions, unless you are an ascetic mendicant.

In many risk analysis approaches on life safety issues, in which the consequences are calculated in numbers of fatalities, even those can be represented by financial numbers. That is despising to philosophers, but practical for engineers. The worth of a human life in risk analysis might be estimated in a range between 3 Million USD [136] and 10 Mio. EUR [104]. In 1990, it had been about 1 Million USD [80]. Life is becoming more expensive – literally. But it strongly depends on the societal and cultural background. Even in so-called egalitarian societies, the efforts to protect VIPs exceed by far the livelihood of a simple peasant. In contrast, in some regions a human life seems to have no value even today.

## 3.2   Health and Diseases

As a simple fact, most people die due to diseases, after having avoided death by accidents, disasters or deliberate killing. Such diseases have not emerged recently, rather we have reduced the risk of many other causes of death that were common in traditional societies. Two hundred years ago, the most probable cause of death in European countries were infectious diseases [15], affecting especially little children.

The situation is similar today in many developing countries. In this respect, hygiene, clean water supply and modern medical treatment are the most essential risk mitigating factors. A breakdown of those provisions, for instance under circumstances of natural disasters and wars, results inevitably in an increase of the disease rate. The resulting casualties most often outnumber those from the immediate killings.

The most probable causes of premature death in modern affluent societies are heart disease and cancer. Few types of cancer are caused by simple cause-effect relations, for instance lung cancer by smoking or exposure to asbestos dust. The most important factor for the risk of getting cancer is age, of course. According to actual research, simple chance might be a major contributing factor in the development of many types of cancer, more influential than any hereditary and environmental factors [134]. Therefore, the cancer rate increases inevitably in modern societies by the fact that in traditional societies, many people died before cancer could break out.

Lifestyle factors leading to disease and early death, like permanent stress, smoking, bad nutrition and physical inactivity, can be seen as an additional exposure to individual risks. However, contrary to widespread belief, a long living population is not necessarily in the interest of society. From an objective – even though inhuman – point of view, the society needs healthy, fit and capable members as long as they fulfill their duties, but then, people who die prematurely save a lot of funds to the public health care system. A healthy population would also thwart the commercial interests of the medical and pharmaceutical industry. The times when medics were predominantly obliged to the wellbeing of their patients are long gone. However, longevity seems to be mainly facilitated by genetic predisposition anyway.

In nature, starvation is a common threat, since food supply is rare and not guaranteed. Energy must not be wasted, and you must take advantage of any of the rare occasions when food is abundant to fill your belly. To cope with that, two major features have developed in most animals and human beings: Laziness and gluttony, particularly the appetite for sweets. 'Cockaigne' had probably been a desirable goal in the Renaissance, but today we have come close to achieve it. By means of agriculture and technology, humans have created a world with abundance of energy and food. Unfortunately, this has also serious disadvantages. Physical activity has been engineered out of the daily lives of most people in affluent societies to satisfy their laziness, leading to inactivity, obesity, and subsequent physical and mental diseases [73].

Fig. 4    Cockaigne *(Bruegel, 1563 - Public domain)*

The transition from hunting and gathering to agriculture thousands of years ago increased significantly the amount of available food, but unfortunately at the price of lower quality. In fact our organism is still adapted to diversified food rather than the one-sided carb diet that consequently leads to many diseases, for instance diabetes. By the way, while overabundance of nutrition is unhealthy for our body, the overflow of information has a similar effect to our mental health. Provisions for moderations and healthy dietary habits in a world of abundance are contradictory to the natural features inherent to us. Religion has declared laziness and gluttony as deadly sins – with little success. Pressure from societal frame conditions and our own idleness prevent us from a healthy lifestyle, which would be no guarantee for health anyway.

Another aspect of risk related to health is suicide. In some societies, suicide is one of the most significant causes of death among young people. Suicide may be seen as the ultimate solution in particular situations, for instance when facing certain death with previous severe suffering. It is an option for those who prefer to choose the moment of their passing, rather than being subject to the power of other people or simple fate. However, in many cases, suicide is attempted by people with a psychiatric illness. Reducing suicide risk is achieved by maintaining mental health, avoiding stress, and giving a meaning to one's life. But that goes beyond the topic of this book.

## 3.3   Where is the Toilet?

On a trip to South America, I traveled from Chile to Bolivia by hitchhiking. The border was situated at 4500 m above sea level on the Altiplano. Since the road was closed for the night, all travellers had to stay in a hut at the customs station. When I felt an urgent need, I asked where the toilet was, and was given the answer: 'outside'. Going outside, I didn't find any toilet, so I did what I had to do behind a bush. Spending the cold night lying on the floor in a tiny room packed with a dozen other men, mostly peasants, truck drivers and a few backpackers, many of them coughing and snoring, I did not sleep very well. The next morning the sun rose on a magnificent landscape. All people who were present, even the foreigners, had to line up in formation when the Bolivian customs officers paraded to their national anthem and hissed their flag. Only after this ceremony, the border was opened. In daylight, I noticed that the ground around the hut was speckled with dots of white toilet paper and small brown heaps. Toilets were obviously unknown, as I had experienced in other traditional societies too, where my demand for a toilet caused only incredulous smiles. A few days after that night, I got high fever, aching limbs and a strong headache, and I had to stay in bed in a dirty hostel in La Paz, Bolivia's capital. My condition was not caused by height sickness, as some colleagues proposed, because I had been quite acclimatized after a few weeks in high altitudes. The disease lasted a few weeks, but instead of curing it properly, I set out to climb Bolivia's highest peak in a weakened state. But that's a different story.

Hygiene is one of the most important safety measures. There is a natural feature inherent in all of us to avoid contact with feces and other excretions, which impose a serious source of diseases. But hygienic conditions vary, and only modern affluent societies, and a few privileged people in less developed countries, can afford a sanitary system with water closets and a sophisticated sewage system. Approximately a third of the world population has no access to proper toilet facilities at all, being exposed to a significant health risk.

## 3.4    Accidents

An accident is a general term for an undesirable, incidental and unplanned event that causes damage and injuries. Such unintentional injuries are the leading cause of death among people below the age of approximately 45 years and the fifth leading cause of death in most developed countries. Generally, people associate the term 'accidents' mainly with traffic accidents. Traffic accidents contribute to a large part of accident casualties, being an unwanted side-effect of a technology that enables us to move with a speed far beyond natural limits, which also applies to other sea, air and land transportation technologies. However, traffic accidents have significantly decreased in developed countries.

Domestic accidents like falling, occurring mostly to elderly people, unintentional poisoning including drugs, and fire incidents, cause most fatalities after traffic accidents. Work accidents have been reduced by stringent accident prevention regulations in most developed countries. Leisure accidents are associated with sports, where risks are taken intentionally. Such accidents were common to our ancestors as well, and therefore could be mostly avoided by the application of natural risk awareness, maintaining of a good state of physical fitness, and appropriate caution. Simply asking 'what could happen' and 'how could it be avoided', the basic risk management questions, is especially important when considering potential accidents. Unfortunately, such risk awareness is impaired in the process of civilization, and people seem to get careless. For young men, risky behavior is natural - going to the limits and beyond is part of the game.

Industrial accidents, as described in this book by many examples, are assigned to technological risks in chapter 3.7.

## 3.5    Natural Disasters

Several times in earth's history, large parts of life on earth were extinct, and each time only a few species had survived. There are many potential causes for mass extinctions, for instance a large asteroid hitting the earth, a supervolcanic eruption, climate change, biochemical processes or extraterrestrial events, like a supernova explosion in the vicinity of the solar system or a gamma ray burst in our galaxy. The probability of such a scenario is considered negligible in the time scale of a human life, but is not equal to zero – it can happen anytime. Finally, the sun will reach the end of its lifespan in a few billion years, and that inevitably will terminate all life on earth anyway.

Approximately 75,000 years ago, a supermassive volcanic eruption occurred at the site of present-day Lake Toba (Indonesia). Volcanic ashes and sulphur dioxide shielding the sunrays off the earth surface led to a global drop in temperature in the following years, which devastated life on earth, and almost wiped out the entire mankind.

Natural disasters in modern human history have been smaller in scale. They are still beyond our control, but in developed societies, we are able to predict such disasters to a certain extent, and significantly reduce the damage. Recent examples of such natural disasters are the tsunami in South East Asia at Christmas 2004, leading to approximately 230,000 deaths, and the earthquake in Haiti on January 12, 2010 with officially confirmed 316,000 fatalities. The resulting damages and numbers of casualties of natural disasters are much lower in developed countries, due to the efficiency of risk mitigation measures, like for instance solid, earthquake-proof constructions, provisions for quick recovery of energy and clean water supply, well equipped and trained emergency services and efficient medical treatment. Such measures have a price, and therefore are not available to those who can't afford it.

In spring 2011, the most powerful earthquake ever registered in Japan occurred, followed by a tsunami that flooded the whole coastal area and killed approximately 18,000 people. Four nuclear power plants in the region were shut down automatically when the quake hit. In one of those plants, Fukushima Daiichi, the flooding caused a complete power supply failure, leading finally to a reactor meltdown and radioactive leakage, but no human fatalities. Paradoxically, Fukushima is perceived in public opinion as the result of technological risk rather than a natural disaster.

## 3.6   Conflict and Violence

In a global and historical context, deliberate destruction and killing by fellow human beings in conflicts and wars lead to the highest number of deaths, right after diseases and natural disasters. Fortunately, violence has declined since the days of our ancestors [105], but the consequences of a war are still completely unforeseeable and can reach an immense magnitude, as the two world wars in the 20th century have shown.

Wars do not only lead to immediate casualties due to fighting, but to famine, diseases and increased exposure to natural disasters due to the breakdown of infrastructure like the supply of water, food, energy and medical care.

For instance, World War I cost more than 17 million deaths on the battlefields and was followed by the worldwide epidemic disease of Spanish Influenza, which led to approximately 25 to 100 million fatalities.

Some of the worst catastrophes in history were in fact war events, for instance:

- The worst fire disaster, probably even the worst massacre in human history in terms of instant carnage, was caused by the bombing of Tokyo by the U.S. Air Force on the night of the 9$^{th}$ March 1945, costing between 80,000 and 130,000 human lives, more than the subsequent nuclear attacks on Hiroshima and Nagasaki.

Fig. 5    Deliberate destruction: Tokyo after air raid in 1945 *(Public domain)*

- The worst maritime disaster was the sinking of the liner 'Wilhelm Gustloff' by a Russian submarine on the 30$^{th}$ January 1945, while evacuating German refugees from Eastern Europe, with approximately 9,000 casualties.
- The worst road tunnel fire in history happened in the Salang tunnel during the war in Afghanistan in 1982 with the number of casualties indicated between 176 and 2,700. The Soviets claimed it was an accident, the Afghan Mujahedeen that it was a deliberate attack.

In books and articles about the importance of alertness in human evolution, it is argued that in prehistoric societies, distracted people would have been killed by beasts of prey, particularly the saber-toothed tiger. That may have been a serious threat, but the most dangerous predators to our ancestors were their fellow humans from the neighboring tribes. Many traditional societies lived in a permanent state of struggles and wars, as show tales from historical sources from ancient cultures, like the Greeks or the Romans, religious books like the Bible and the Quran, or anthropologic literature, for instance Diamond [32]. Archeological relicts show surprisingly many traces of violent impacts on bones and skulls. A famous example is 'Ötzi the Iceman', the prehistoric hunter that had been conserved under an Alpine glacier for more than 5000 years before he was released in 1991 due to climate warming. His body showed scars, and an arrowhead stuck in his shoulder. Very probably he had been shot while having a rest on a journey over the mountains [82].

Conflict and violence are natural features of human interaction. The risk of being killed by your fellow humans does not only arise due to conflict with other groups, but may be inherent in a society and a state, particularly under ruthless totalitarian authorities. Shortage of resources, poverty and social inequality lead people to apply violence as a means to improve their situation. For them, the chances are simply worth the risk. Unfortunately, with the development of technology, the possible consequences, and therefore the risks, have risen to an uncontrollable level. With the introduction of modern weapons of mass extinction, men have got an unprecedented means of extermination. After a few decades of peace and security in West Europe, North America and East Asia, such risks are strongly neglected.

Willful damage and destruction must be considered as a possible scenario in any risk consideration. Technical facilities with inherent risks, like transportation infrastructure, power plants or chemical production sites, where an appropriate level of safety is achieved only with extensive operational and technical measures, may become key targets in a conflict. On a small scale, people and facilities may be protected from an attack and malicious damage by security measures like surveillance systems, access control and the employment of guardians. Security works mainly by deterrence, increasing the price that would have to be paid by an opponent to achieve his goal.

As with most risk issues, the principles to mitigate violence risks are simple in theory, but complex in reality. Going into the details would exceed the topic of this book. Effective conflict risk management might be described by an African proverb that became famous by U.S. president Theodore Roosevelt (1858 – 1919): 'Speak softly, but carry a big stick'.

## 3.7 Technology

In public discussions, risks related to constructions and technologies are given extraordinary importance. People are more afraid of technological risks than of those resulting from natural disasters or diseases. However, technology has been invented in the first place because of its advantages. Any technology may serve its purpose in a given context, but we rarely know about all of its possible benefits and risks. Risks result form inevitable drawbacks, unwanted side effects and possible failures.

Such an obvious unwanted side effect is the air pollution caused by furnaces and combustion engines, leading to massive environmental disasters, for instance the Great Smog in London in December 1952, causing thousands of fatalities. In the late 1980s, just before the fall of the communist dictatorship, I could finally travel to my Czech home country, after 20 years of exile in Switzerland, to visit my relatives. Pampered by mountain air, I could hardly breathe in the prevailing smog, which resulted from combustion of lignite in power stations and furnaces. The situation is similar today in developing countries, whereas in most affluent, industrialized countries, the pollution of air and water has been significantly reduced in the past decades due to public awareness and stringent environmental protection legislation.

As part of the development process, technologies become safer, cleaner and more reliable. The safer a technology is, the more important the residual risk becomes. Failures may lead to the non-fulfillment of the primary purpose, or even to additional risks. Consequences must be determined for possible failure scenarios. Which measures need to be taken in order to ensure the required level of safety, depending on the distinction between systemic and coincidental failures? Humans associate technologies with great benefits to low risks, and vice versa, based on risk perception by feelings [122]. Reality is different. Technologies with high benefits may be accompanied by high risks. Further, even the safest, theoretically perfectly reliable technical facility is prone to deliberate damage by hostile human beings, or by unexpected natural disasters.

The most obvious criterion is whether it works. But 'working' may be the consequence of temporary luck, rather than the real suitability of underlying concepts and technologies. Whether it really works, you see only when you stress it under various conditions. This aspect will be explained later in detail.

## 3.8  Tōhoku Earthquake and Fukushima

On March 11, 2011 the most powerful ever registered earthquake in Japan and globally the fourth most powerful in human history occurred off the Japanese East Coast. Subsequent tsunami waves inundated 561 square kilometers and left approximately 18,000 dead and many still missing. Following the earthquake, incidents occurred in various Japanese nuclear power plants. Eleven reactors had to be shut down. At the two facilities Fukushima Daiichi (I) and Daini (II), tsunami waves with heights exceeding 10 Meters overtopped seawalls, flooded the buildings and destroyed equipment. At Fukushima Daiichi, this included the diesel backup power systems that were required for emergency cooling. As a consequence, three of the six reactors suffered nuclear meltdowns, leading to explosions and severe radioactive leakage [55].

Fig. 6     Tsunami in Japan on March 11, 2011 *(AFP/Gettyimages)*

At Fukushima Daini, situated approximately 12 km south of Fukushima Daiichi, the four reactors stayed intact, despite temporarily insufficient cooling of three reactors. In contrast to Fukushima Daiichi, the off-site power at the Daini facility remained at least partly functional and damages could be fixed quickly, whereas the Daiichi plant suffered a complete loss of power supply.

The total amount of radioactive material released to the environment was much lower than in Chernobyl 1986. There were no immediate fatalities, even when some station workers received high doses of radiation. In Fukushima Daini, a worker died in an accident during the earthquake, which had nothing to do with the subsequent nuclear incident.

In fact, the Fukushima Nuclear Disaster was only a minor secondary effect of a major natural disaster. Earthquakes occur frequently in Japan. The measures against earthquakes proved their worth, but those against tsunamis were obviously underdesigned. At Fukushima Daiichi, all levels of redundancy have failed, particularly the auxiliary power supply. Secondary measures, limiting the extent of damage, were difficult to deploy, taking into account that the whole infrastructure had been destroyed by the tsunami.

Fig. 7    Destroyed reactors at Fukushima Daiichi *(Creative Commons)*

The Fukushima Nuclear Disaster gives a vivid example on the limits of risk analysis. When the plant was built in the early 1970s, the design basis tsunami height had been 3.1 m. This value was based on historical tsunami records. In 2002, the design basis was reassessed using deterministic methods and the height increased to 5.7 m. On March 11, 2011, the waves topped twice that value. In this respect, the Fukushima Nuclear Disaster can be seen as an example of insufficient safety margins. However, that's the hindsight view. Before 2011, it might have been difficult to explain to investors why to build protection walls exceeding three times the height of historical tsunami records.

One positive aspect has to be pointed out. Despite the complete destruction of the most necessary equipment by the tsunami, the engineers and workers in Fukushima achieved to stabilize the situation, restoring improvised power supply, managing successfully a controlled reactor shutdown in Fukushima Daini, and preventing further overheating and limiting the release of radioactivity as far as possible in Fukushima Daiichi.

They had to work under extreme stress and uncertainty, threatened by subsequent earthquake aftershocks and tsunami alerts, without appropriate information about the actual state of the plant, facing an unprecedented challenge. Those professionals owe our highest respect!

Beside Fukushima Daiichi, all other Japanese nuclear power plants, some of them also heavily affected by the unique earthquake and tsunami, endured that catastrophe without nuclear damage. The fear generated by that incident caused a global anti-nuclear backlash, which is difficult to explain when considering the simple facts. Unlike the preceding natural disaster, the incident in Fukushima Daiichi had neither killed nor injured any people. The radioactive contamination was severe, but limited. In contrast, the evacuation of approximately 100,000 residents, as mandated by the government, caused various casualties. The ecological disaster following the destruction of other industrial plants by the tsunami, like oil refineries, has also been overlooked by the overstatement of radioactive contamination.

## 3.9    Depletion of Resources

Safety has a price, and this can only be paid in a wealthy society. Our wealth is dependent on natural resources that are about to be depleted sooner or later, for instance fossil fuels, while the population in demand of such resources is growing on a global scale. As history has shown, such a development will inevitably end up in a breakdown.

Both traditional and modern societies neglect the risks deriving from the lack of vital means of livelihood. Many cultures have vanished after they ran out of natural resources, in particular fresh water and food supply, due to depletion or climate change [30]. Shortage of supply has always been a problem to single individuals and groups, but as long as the effect was locally limited, it never affected the whole population. On the other hand, overpopulation is not a problem in a world of abundance. Simply put, supply and demand must be in accordance. In the past decades, the increase of supply of food and energy has led to an enormous global population growth. Unfortunately, those supplies are not guaranteed in the long run. For individuals, overpopulation leads to a miserable life in poverty, which is the case for many people on earth today. As a collective risk, overpopulation may end in a catastrophe when the supply of basic vital resources like clean water, food and energy will be depleted. However, this may occur not as a singular doomsday scenario, but rather as a slow process, which is difficult to perceive from the limited point of view of a human life.

It would be wise to reduce the demand before supply runs short. Shortage of resources is the fundamental cause of many wars, which impose one of the highest risks to humans in a global, historical context. Our ancestors applied the principle of limiting demand during most of mankind's history. The obvious solutions to those problems are simple in theory:

- Limiting human population
- Limiting energy consumption and material demands
- Avoiding harmful emissions
- Finding new resources
- Transformation to a fully sustainable economy

In practice, the problem of overpopulation is a delicate matter, and cannot be subject to a simple and fast solution, since our ethical values and appreciation of human life in a liberal society must not be questioned. Further, our psychological and societal boundary conditions prevent us from doing the right thing. The implementation of the necessary measures is not popular, and any politician that will propose so will have a hard stand. Such is even worse in traditional societies that are based on superstition and religious beliefs, but use modern technology and medical achievements that enable uncontrolled proliferation.

Fig. 8    Famine in Africa *(Public domain)*

# 4    Historical Aspects

## 4.1    Origins

Human risk perception and behavior can be better understood when recalling for what kind of risks it was developed. However, we need to take into account that history is mostly based on assumptions and incomplete information. Historical records are biased, and archeological evidence is limited. In principle, we can never know what it really was like in the past from our present point of view. Humans might not always have had the same set of faculties and thought processes as today. Statements about the past, like in the following chapter, always must be treated with a healthy dose of skepticism.

Our species – Homo Sapiens – has been around for at least 200,000 years. Our predecessors used stone tools already 2.8 million years ago. Most of the time, humans lived at a low population density in small egalitarian groups of hunters and gatherers. Such 'original' societies based their livelihood on a diversity of natural resources, like edible plants, roots, fruits, nuts, small animals and occasionally big game, and most importantly, water. Process energy was provided by firewood. Starving was a regular threat, particularly in winter. Strict birth control was essential, to avoid exceeding the size of population that could be maintained. The demand for mobility limited the number of little children that had to be carried. Competition for resources resulted in violent conflicts between different groups and individuals. Unexpected natural disasters happened from time to time. Those conditions prevailed for the vast part of human history and have shaped our genetic disposition and social behavior.

## 4.2    Traditional Societies

Humans have developed awareness for those traditional risks in an evolutionary process over millions of years. Being focused mainly on situational awareness, avoidance of known hazards and fast response to immediate threats, our ancestors have survived, proliferated and spread all over the world, having restrained many dangers. Simple rules took into account the experience of many generations in an almost constant, unchanging environment. Following traditions was usually a good approach. Probably they had no idea about statistics and probabilities, neither had they access to the necessary data and appropriate means of analysis and calculation. Men thought that they had little influence on the future and its risks, and fate was determined by unswayable forces or by the Gods.

Many anthropologists have investigated risk management in traditional socie-ties, for instance Diamond [32]. The attentiveness of traditional hunters and other people that live in nature is astonishing. Smallest details in the environ-ment are noticed and assessed. Reactions to immediate threats, like 'freeze, fight or flight' are carried out instantly and mostly unconsciously. Unfortunately, in modern societies we haven't got the same ability for instant perception and assessment when we face a critical situation as a responsible operator of a dangerous technology, receiving the relevant – and mostly not so relevant – information from a computer screen.

In traditional societies, abstract risks with low probabilities of occurrence and time delay between cause and effect are neglected. For instance, the key role of hygiene in preventing infectious diseases has been discovered only recently in the 19th century. Risks to the environment, which are ubiquitous in the as-sessment of new technologies, are rarely recognized, simply because the envi-ronment is perceived as unlimited. When an area was depleted, you moved to another, unspoiled place. The self-purification capacity of water and air was considered to be inexhaustible. However, with increasing population density, unspoiled areas became rare, and contaminated water is still a major source of diseases.

## 4.3    Adapting Agriculture and Technology

Some human societies made a crucial decision approximately 10,000 years ago when giving up the life as hunters and gatherers, adapting to a life as farmers in a slow process. Agriculture started to develop. Instead of controlling demand, they increased supply. The evolution of today's human culture and technology wouldn't have been possible without that step. Nevertheless, for most individuals, this transition to guaranteed supply in quantity led to many disadvantages. The step from hunters and gatherers to agriculture led to a massive population growth, accompanied by poor nutrition in quality, increased contagious diseases, and social inequality, leading to exploitation and discrimi-nation. Such issues are still common in many developing countries. In fact, until the end of the 18th century, for the majority of people living in agricultural cultures, quality of life in general seemed to be worse than in societies of hunt-ers and gatherers [25]. Only a few percent on the top of the social hierarchy could afford a higher standard of living.

Why did agriculture prevail? The driving force of human development seems to be rivalry and conflict between different individuals and groups. In large chiefdoms and the first states, law and order was enforced, thus significantly reducing the violence risk within the group. Those groups gained superiority in wars with their weaker neighbors. The increased population, fed with abundant food supply, improved strength by sheer numbers. Bread feeds cannon fodder. Furthermore, agricultural societies also led to division of labor and a massive technological advancement, providing superiority in almost all relevant aspects in conflicts. The remaining hunters and gatherers were wiped out, assimilated or pushed to poor habitats.

## 4.4    Enlightenment

Due to increasing scientific knowledge during the period of the Renaissance and Enlightenment, mankind has developed concepts of 'probability' and 'risk', instead of simply taking events as granted. This was one of the most important shifts in paradigm for human culture. Fate seemed to become calculable, although not predictable. Men learned to arrange their future, thus taking responsibility into their own hands, and the Gods were left behind [13]. The need for risk calculations first arose from the necessity to protect trade against unforeseen losses. This was particularly important for ship transport, which was exposed to the unpredictability of the rough sea. Many ships and their freight got lost in storms or pirate attacks. To provide reimbursement, insurance companies were founded. To be able to calculate reasonable premiums, analysis of past incidents had to be worked out, leading to estimates of possible losses and approximate frequencies of occurrence. Another need for estimation of probabilities came from gambling, which was a favorite pastime throughout history. Unlike in real life, the odds in gambling, for instance in the game of dice, follow strictly simple mathematical rules, providing a good basis for the development of probability calculus. Therefore, professional risk management was first restricted to insurance and gambling. Abstract risks could not be handled.

In traditional societies, from the point of view of a lifespan of a single individual, progress could hardly be perceived. People lived in a steady environment, with known dangers, that were handled by traditional risk management. There was rarely something new to care about. Technological risks became an issue by the fast scientific and technological progress after the Enlightenment and the Industrial Revolution, which was mainly enabled by a massive increase of availability of energy, based on fossil coal.

New civilization diseases, means of modern warfare, and risks inherent in new technologies, particularly those providing energy, means of transport or communication, have arisen in the past decades. Those are generally more complex and less obvious than the risks that our ancestors have faced. In traditional societies, there is a strong lack of prudence considering such modern risks, as can be seen for instance in the high road accident rate in developing countries. There was no time to develop appropriate risk awareness. In this respect, traditional risk management with its shortfalls must be replaced by scientific progress and better insight. Alfred Holt (British engineer and merchant, 1829 – 1911) wrote in 1877 [53]:

*'It is found that anything that can go wrong ... generally does go wrong sooner or later, so it is not to be wondered that owners prefer the safe to the scientific ... Sufficient stress can hardly be laid on the advantages of simplicity. The human factor cannot be safely neglected in planning machinery.'*

140 years ago, the basic principles of risk management were obviously not so different from today. A conservative approach was pursued, since people don't like drastic changes – unless their lives have become intolerable. However, caring about technological risks is predominantly only an issue in affluent societies. In developed countries, we have overcome many traditional risks by scientific and technical progress, which we take for granted. Any further improvement is not basically necessary, and the accompanying risks are questioned. In contrast, in developing societies, almost every technological innovation does increase prosperity and safety in the first place, at least to some. Questioning the risks of such technologies is not feasible.

On the other hand, the calculability of risks by modern risk management may have been overestimated, and the influence of mere chance is higher than most people think. Risk management implies a free will and the means to plan and influence the future. Unfortunately, human behavior is unpredictable in principle. Even the free will of humans might be questioned, at least according to neurological research [81].

In natural science, certainty and predictability do not exist. The mechanistic conception of the world, which was one of the basic sources of technological progress in the past 300 years, has come to an end due to the new insights from quantum physics and chaos theory in the 20th century.

## 4.5 In the Jungle

I learnt some valuable lessons about traditional risk management on a journey to South America in the early 1990s. After having travelled for many months and climbed some peaks, I got an invitation to the Bolivian lowlands, at the edge of the Amazonian jungle. With my mountaineering equipment, I was completely unprepared for the environment where I was about to go.

Fig. 9    Camino de las Yungas, the 'death road' *(Creative Commons)*

The journey from the Bolivian highlands to the lowlands by the 'Camino de las Yungas', which by then was recognized as one of the world's most dangerous roads[6], was an adventure by itself. This single-lane road winds up in steep serpentines along high cliffs, and was passed by heavy trucks and overloaded busses on a dirt track without guardrails. Despite the obvious danger, most drivers seemed to be quite careless, neglecting road safety rules by 'western standards'. Noticing the many burial crosses along the roadside did not really inspire my confidence. Luckily, we reached the final station of the bus, and after another trip down the Beni river on a canoe I arrived at my host's village.

---

[6] With an average of approximately 200 – 300 fatalities per year on a length of 65 km, until 2007 when a safer bypass road was opened.

The native Tacana people lived on the edge between tradition and modern civilization as farmers, hunters and fishermen, wearing modern clothes, using flashlights, firearms, and diesel engines, and participating in Lutheran sermons provided by Swiss missionaries. After a week of socializing and preparation, we left for a hunting trip further down the river, to the jungle of the Amazon basin. That environment was absolutely unknown to me, and my experience as a mountaineer, as well as my previously achieved MSc. in engineering, were of very limited use. The risks in the jungle arise not only from wild beasts, poisonous snakes and dangerous insects that you would expect anyway, but from simple dangers like getting lost in a completely flat, wooded area. In this respect, the sense of orientation of the natives is legendary. Some people say they can smell the direction. In fact, it is not about smelling, but simple orientation tricks, which they use almost unconsciously. Another unexpected risk in the rainforest is to suffer from thirst and diarrhea. Apart from the rivers and pools, water was to be found only in muddy pits. The primary source of fresh water were liana. Unfortunately, some were toxic, and completely undistinguishable to my eyes.

In the 'civilized' world we are used to medical care and trauma surgery, and sometimes quite careless about risks of possible accidents. In the wilderness, such carelessness can be fatal. The Tacana people were very cautious, avoiding danger whenever possible. They would have little comprehension for thrilling sports in modern societies. On the other hand, occasionally they risked their neck when there was an opportunity to catch an extraordinary prey, whether on a hunt or on a fishing trip. That would be possibly not understandable for people who are used to buy their food at the supermarket whenever they are hungry. For the Tacana people, that was part of earning their living.

Some inconveniences were simply neglected. For example, the level of the river varied several meters within a few days, depending on the rain and melt water from the far Andean mountains. The natives took into consideration that their canoes could sink occasionally, when they were tied too close to the ground and the water level dropped and afterwards rose strongly overnight. That was no problem for a traditional canoe, you just had to turn it, pull it out of the water, and let it dry. One morning I came to the shore and all canoes were under water, but one of those was equipped with a diesel engine, which was then flooded too. This mishap could have been avoided by simply tying the canoe to a longer rope, but obviously this was not an issue of consideration. The owner dismantled the engine (spilling the motor oil to the soil), dried it and made it work again. That was performed by using just a few simple tools, and the repair process took almost the whole day, employing three men who were mainly hanging around, discussing and commenting.

The Tacana people had an incredible nature awareness, an immense memory capacity and sometimes bright ideas. An important aspect is that modern media did not spoil their consciousness. There simply were almost no such media, beside some occasional TV, supplied by a diesel generator, in the larger villages. The most common source of information was oral tradition. That does not mean that there would not be occasional wrong risk assessment. Sometimes they narrated fairytales with significant exaggerations, but those were usually not about risks that directly threatened them in their daily life. On the other hand, the Tacana people were quite careless about risks resulting from modern technologies and environmental hazards. Appropriate risk awareness could not have developed in the short time since those technologies were introduced, and they did not use written information, which is essential for us. As an example, all the rubbish, and even cadavers, were thrown into the same river, from which they obtained water for cooking, washing and drinking. Apart from some tapeworm that came out a few weeks later, I remained relatively healthy despite drinking the same water. However, when my host's wife threw old batteries into the water, I tried to explain that batteries contain poison, but that was not heard. The native people relied on the self purification capacity of the water, which had obviously always worked.

Venomous animals were one of the threats that I feared most. My hosts warned me particularly about some of the snakes, and since I was not able to distinguish them, I considered them all as dangerous. Once I walked barefooted through the grass, when suddenly a tiny little snake appeared in my path. I cut its head off immediately with my machete, and showed it to my companion. He confirmed that is was a poisonous species. I realized how lucky I had been, because usually my alertness was far below that of the natives. The snake was not so lucky, and its cadaver was fed to the chicken. Worse than the snakes were the many species of insects. I remember a kind of ant that occasionally jumped from the trees on my head, biting and causing an enormous pain. After this had happened a few times, I got paranoid about that. Even worse were the tiny insects that laid their eggs under the skin of my legs. I could get rid of them only months later, after having returned to civilization. Obviously those vermin preferred the guest from Europe to the locals, even when many of the natives were also full of infected scars from such bites.

I also learned a lesson about attitude. My host, an old and wise, but physically extraordinarily fit man, reported that in his childhood, they were still hunting with bow and arrow, and very rarely a white man came around. One generation later, his children used firearms and flashlights.

He narrated a story of a white man who had died after a bite from a poisonous snake. Being extraordinarily stressed in the face of approaching death, he collapsed due to a circulatory breakdown. In contrast, the natives often survived even such potentially fatal snakebites. When they realized they were bitten, they simply accepted the fact that they were about to die. There was no medical help to be expected anyway, so they sat down and waited quietly for death to come. Often, their organism managed to overcome the poisoning, but it required patience, strong health and the ability to suffer. Their attitude of disregarding near death helped them to survive. The most important lesson was that the native people were not as stressed as citizens of modern societies. The state of permanent activity and exhaustion, as observed in most professional environments in developed countries, is unthinkable. A lot of time is spent simply waiting and observing, the same way as I do while spending time in nature. From our economic point of view, this may be highly inefficient, but it provides reserves that are essential to be prepared for risks and chances of unforeseen events.

Fig. 10     In the jungle

# 5 The Scientific Approach

## 5.1 Risk Management

Risk management is common practice in any engineering process of technical systems with inherent risks. The tools are quite sophisticated, and there are many professionals handling successfully the risks in their specific field of activity. In this respect, industrial safety culture led to a constant improvement in the past decades. One of the most important driving forces in the development of modern risk analysis was nuclear-power reactor safety.

The process of dealing with risks is defined in many applicable rules and guidelines. According to the guide [59], risk assessment is that part of risk management which provides a structured process that identifies how objectives may be affected, and analyses the risk in terms of consequences and their probabilities before deciding on whether further attention and treatment is required.

Risk management attempts to answer the following fundamental questions:

- What can happen and why (by risk identification)?
- What are the consequences?
- What is the probability of their future occurrence?
- Which factors reduce the probability of the risk?
- Which factors mitigate the consequence of the risk?
- Is the level of risk acceptable, or does it require further attention?

A lot of literature is devoted to this matter, and for many professionals, risk management is daily routine. Usually, a simplified probability-consequence chart similar to Fig. 11 is proposed.

| Conse-quence | high | transition area | not acceptable | not acceptable |
|---|---|---|---|---|
| | medium | acceptable | transition area | not acceptable |
| | low | acceptable | acceptable | transition area |
| | | low | medium | high |

**Probability**

Fig. 11    Risk management in theory: Simplified probability-consequence chart

Quantitative Risk Analysis (QRA) is a standard approach [10] based on evaluating and calculating the probabilities and consequences of all relevant individual scenarios. Any risk can be seen as a chain of multiple elements, each one with a certain probability and consequence. The risk of a single event is defined as the product of its probability and the resulting consequence. The total risk in a given context is the sum of risks of all possible events. Multiple events may be evaluated by an event tree analysis, where some single initial events lead to multiple subsequent events, each of those leads to a number of further events and so on. Most importantly, high-consequence scenarios may emerge as a result of the occurrence of multiple high-probability and low consequence events. The calculated distribution of risks with different probabilities and consequences can be displayed in a frequency-consequence-diagram (F/C-diagram)[7], like for instance in Fig. 15 on page 48.

The risk value provided by QRA must not be interpreted as an absolute reference. QRA serve rather as a uniform comparison standard for the evaluation of different conceptual and technical options, referring to a reference case. By that, variants of safety concepts and equipment can be compared within defined system boundaries and assumptions. The benefit of a concept is quantifiable, for example, in the reduced death count for each analyzed option in comparison to an alternate option. The definition of a basic concept, serving as a reference against which all other options are compared, is essential, but always rather arbitrary[8]. By that, we get a perfect tool for decision making in theory. Unfortunately, practice is different.

## 5.2 The Worst Case

*'Whatever can go wrong will go wrong'*
(Edward A. Murphy, Jr., U.S. aerospace engineer, 1918 – 1990)

Murphy's law defines in a clear and simple form of what has to be expected in real life. An often-used term in that aspect is 'worst case'. As a matter of principle, this term is nonsense, if not assigned to a defined context, within a specific period and fixed boundary conditions.

---

[7] Here, the 'frequency' is the reciprocal value of the probability of a risk, and does not refer to periodically occurring events. An event that happens with a mean frequency of once in a million years can occur tomorrow.

[8] For instance, according to most road tunnel QRA standards, a concept according to prescriptive guidelines has to be taken as a reference, even when it rarely represents a good solution.

*'There is no worst case. There can always be a worse one!'*

A proposal for such a case would be the extinction of all life on this planet by a global disaster. However, that is not really helpful in any risk assessment. The question of what can happen may be answered by experience from past events and by using your imagination to invent useful scenarios. When engineers use the term 'worst case', they in fact refer to a 'design case'. For most applications, defining appropriate design cases is useful and necessary, especially when the consequence of a single event is physically limited and simultaneous multiple events are not taken into account. A design case defines the scenarios that are taken into account. In other words, events with a possible damage higher than the design case are left out and perceived as acceptable residual risks. Usually those are events that are very unlikely, and have never happened before. When the worst case that has happened so far is determined as a design case for safety measures, then any worse case may lead to a disaster. An illustrative example for unforeseen exceedance of design cases are the seawalls in Fukushima.

In fact, knowledge obtained from past experience and inductive reasoning is not a guarantee for safety. This is known as the problem of induction, formulated by David Hume (Scottish philosopher, 1711 – 1776). An illustration of this problem is a chicken on a farm, as proposed by Bertrand Russell (British Philosopher and Mathematician, 1872 - 1970). The farmer feeds it every day, the chicken is happy and assumes that this will continue indefinitely. From the chicken's point of view, it's the only logical assumption. But one day, the chicken has its neck wrung, and is processed to soup. Its death was unexpected – to the chicken, not to the farmer.

A systematic approach to risk scenarios of technical systems is based on cause analysis, addressing inapt concepts, design flaws, mistakes in the implementation, human errors, or unfavorable boundary conditions. Natural disasters may be considered as extremely unfavorable boundary conditions. Another aspect are risk mitigation measures. Such measures might have unwanted secondary effects, some even increasing the risk. What happens when the safety equipment fails? What happens in case of an erroneous release? For instance, an accidental release of sprinklers in a road tunnel under traffic may cause serious accidents – which in fact impose a higher risk to the road users than the fires, which the sprinklers are supposed to suppress.

When analyzing risks, the possible events to be taken into account for a given context have to be defined and limited, even when the number of possible events is literally countless. The fundamental problem is that by this limitation, an unknown part of the risk is neglected. The question 'what can happen' can never be fully answered, and some risks stay unknown by principle if the disasters doesn't happen.

## 5.3    Consequences

Since risk is the product of probability and consequence, the consequence must be quantified. The understanding of what has happened or what might happen is essential for evaluating the consequence in any risk consideration. This can be modeled either by tests or by simulations. Any design calculation can be seen as a simplified model. The extent of damage of a scenario can be calculated by simulation models or measured during tests, for each considered event. By means of computer simulations, by varying initial and boundary conditions and other determining parameters, a huge number of scenarios can be simulated. Such an approach is particularly useful in particular for complex systems, where the possibilities of failures of risk mitigating measures should be considered too. However, predefined scenarios never have the full depth and width of uncertainty in reality. Even when taking into account all secondary effects and consequential damages of even simple scenarios, the total damage cannot be thoroughly calculated. Any resulting number is a mere estimation. However, the same is true for benefits. Often you can profit from unforeseen aspects that you haven't thought about previously.

The possibilities for simulations are limited, in particular when the models for calculation of damages are complex. In many cases, it would take years to simulate all reasonable scenarios, and an eternity to simulate 'all possible' scenarios. Therefore, the number of scenarios has to be severely restricted, hopefully to some significant cases, which have to be defined based on assumptions. The trustworthiness of models is always questionable. Computer simulations lead to a progress in the development of new technologies. Simulations can be used as a shortcut in the development process to the final product, but the simulation must not be the final step. Based on the findings from the simulations, a construction or a prototype may be built, which can be tested in reality. By simulations, the process of testing can be significantly reduced, but not omitted. Simulations cannot replace subsequent testing. Many aspects that have not been addressed in the simulations would be discovered in tests and corrected before the operation starts.

Events with high consequence cannot be reasonably tested. Therefore, modeling and simulating disasters, for instance fires or explosions, based on validated physical models and reasonable input data, is an important tool in risk analysis[9].

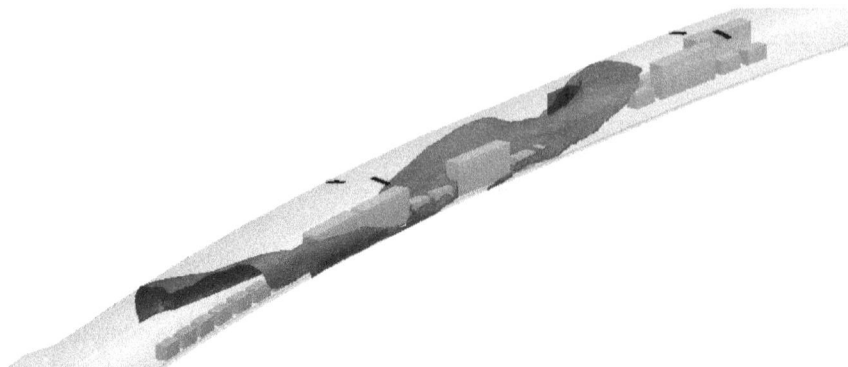

Fig. 12    CFD simulation of smoke spread resulting from a fire in a tunnel

Serious simulation models are approved through similar tests and measurements, but the simulated scenario depends not only on the model, but also on boundary and initial conditions. Such conditions are always based on assumptions, thus comprising a possible subjective error. In fact, the boundary conditions may be the most decisive factor for the modeled extent of damage, for instance in simulations of smoke propagation for fire safety applications. Fire simulations that don't take into account the natural flow and meteorological conditions are surprisingly common, but questionable.

Many academics and theorists rely too much on modeling and simulations. A model must never be confounded with reality. Simulations of events that (hopefully) never happen in reality and therefore cannot be observed and measured bear an inherent uncertainty. Real incidents, models and test scenarios are three different issues; the distinction between them is essential. In simulations, there is never the same level of confidence as in a series of real tests. On the other hand, when a risk analysis, based on modeling for many different scenarios, is in accordance with experience from real incidents, then confidence might be justified, on the premise that data from a significant number of incidents are available.

---

[9] E.g. CFD (Computational Fluid Dynamics) simulations with physical aero- and thermodynamic modeling of smoke spread in fire scenarios, which may take several hours or days of computation time for each scenario.

For instance, risk analysis for road tunnels is based on many fire scenarios with simulation of smoke spread and personal escape for different boundary conditions and parameters. Most scenarios result in no fatalities at all, but a few scenarios, sometimes only one single incident, may have catastrophic consequences with many casualties. Such findings correspond to what can be observed in reality. For calculating consequences, subsequent secondary effects have to be taken into account. A carefully analyzed example for such a secondary effect is the significant increase in road traffic fatalities in the U.S. after 9/11 [45], when people were afraid of flying and preferred to drive by car, neglecting that the risk of road traffic is generally higher than that of flying.

As a final remark, bear in mind that thorough analysis and decision making require time and energy. Being concerned with risks with low consequence, you lack resources for really important decisions. If the consequence of an identified risk is bearable, any further risk considerations would be of little use. Simply accept the risk and take the possible damage into account.

## 5.4   Probability and Predictability

'*The theory of probabilities is basically just common sense reduced to calculus*' (Pierre-Simon Laplace, French mathematician and physicist, 1749 – 1827)

Risks are about probabilities, not about certainty. There always remains an element of simple chance, regardless of risk mitigation efforts. Unfortunately, as humans we don't really understand the term 'probability'. A distinction between deterministic and statistic quantities is essential. Cause and effect relationships are deterministic. The probability of something that has already happened is by definition 100%, but that insight is useless. What counts is the probability before it happens. In contrast, a random process with an element of chance and indeterminacy is stochastic. The result is not predetermined, and therefore in principle not predictable. Something might happen or not, which brings with it an element of uncertainty in principle. We can only try to evaluate probabilities of different outcomes.

When driving a car into a massive concrete wall exceeding a speed around 100 km/h, dead is practically certain. The impact energy is too high to leave any chance of survival. That's why for instance lateral walls in niches of highway tunnels provide a source of significant risk.

However, by changing boundary conditions, for instance by reducing speed or replacing the concrete wall by soft panels, the chance to survive is increased. At a sufficiently low speed, around a few km/h, damage is reduced to an extent of almost guaranteed survival. Nevertheless, you still might die of a heart attack when startled by the impact.

We tend to see most effects in the world as linear and deterministic, therefore predictable, which is only a rough simplification. Determinacy as such does not exist in principle, as is implied by Quantum Theory. Physical systems seem to behave deterministically only on a large scale, but on the smallest scale they are ruled by randomness. Even deterministic systems can become unpredictable, due to nonlinear relations, as is described by Chaos Theory. The dynamic behavior of even a very simple system, for instance a double pendulum, may be unpredictable. The classic example for a nonlinear system is the weather. A butterfly in Asia can cause a hurricane in the U.S. However, nonlinear systems do not necessarily lead to uncontrollable disasters, but may rather converge towards a stable, steady state, or stay in an oscillation between boundary limits. The extermination of butterflies in order to reduce the risk of extreme weather phenomena would be a bad idea, since the achievement of that goal is rather questionable, and might lead to much worse consequences. The vast majority of butterflies don't cause hurricanes, but they have probably an important role in their ecosystem – even when that role cannot be easily recognized. In the case of butterflies this is obvious, but what about other risk issues where knowledge of probabilities and consequences is largely overestimated?

Fig. 13    Has it caused a disastrous hurricane? *(Creative Commons)*

Many health issues show stochastic characteristics. For instance, the cause-effect relation between smoking and lung cancer was discovered in the 1950ies. According to statistics, the probability to die of lung cancer is approximately ten times higher for a smoker than for a non-smoker, but you may spend your whole life smoking cigarettes without getting lung cancer, even when possibly suffering from other diseases. In contrast, some people may get lung cancer without smoking. That's the favorite argument provided by smokers to defend their bad habit, but in a stochastic process, single examples are not useful to prove anything. Probabilities may be determined based on statistical data available from previous incidents, in particular for events that happen regularly. Such data for significant individual risks are available on the internet from many public sources, as in chapter 14.2. Consider the following probabilities of causes of death, applying to most developed, affluent countries:

- Since the probability of death is 100%, age is the major factor influencing the probability to die. Just as a reminder, risk as referred to in this book is about dying 'prematurely due to an avoidable cause'.
- Approximately 20 - 30% of all deaths result from potentially avoidable causes, however this is practically difficult to estimate.
- The majority of people die due to diseases, mainly by heart disease (approx. 30%) and cancer (approx. 20%).
- For people below 65 years of age, the most probable cause of death is cancer.
- For people below 45 years of age, the most probable cause of death are accidents, contributing to approx. 5% of all deaths.
- Road traffic in the U.S contributes to approx. 0,9% of deaths.
- Homicide in the U.S. contributes to approx. 0,4% of deaths[10].
- Road traffic in Switzerland contributes to approx. 0,34% of deaths.
- Homicide in Switzerland contributes to approx. 0,06% of deaths.
- The probability to die due to an earth-impacting asteroid is in the range of approx. 0,000002% (2.E-8) [23][11]
- The chance to win the lottery is approx. 0,000006% (6.E-8)[12]

---

[10] The U.S. are very heterogeneous in that aspect, being strongly dependent on the local and sociocultural background. For instance, the homicide rate in Louisiana is more than 10 times higher than in New Hampshire.

[11] In contrast to the other mentioned probabilities, this one cannot be based on statistical evidence, and therefore is only a rough estimation based on rough assumptions.

[12] Example for one pick 6 from 50 numbers game, which can be exactly calculated, based on fixed and simple mathematical conditions, in contrast to real life events.

These are average values over the whole population in developed countries, to be used as an approximate benchmark magnitude to judge risks that are calculated by risk analysis in numbers of fatality rates. When the estimated probability to die due to accidents or failures of technical systems is in the same magnitude as the (negligible) 'mass extinction by asteroid scenario', does it make sense to care about it?[13] Taking into account probabilities, preferably based on trustworthy statistical data, is a good cure against availability and representativeness bias. However, the significance of the mentioned probabilities must not be overestimated. Statistics do not represent reality. They are only a mathematical model to approach its understanding. In real life, risk factors are often unknown, combined and not independent from each other. For a particular individual, the probability of any cause of death depends on age, gender, domicile, occupation, lifestyle and many other environmental and social factors.

For many risks, in particular those with high consequence, the statistical data are not significant enough, i.e. the number of events is too low – fortunately! In that case, probabilities may alternatively be calculated based on models, strongly affected by assumptions on the input data. Probabilities are often displayed as frequencies, as for instance according to the OECD/PIARC methodology [103]. Unfortunately, this leads to serious misunderstandings. Frequency is usually assigned to anything that happens regularly. In contrast, probability is a statistical value: Even highly improbable events with a low average frequency may happen just the next moment. That's what people hope for when playing lotto!

For instance, I worked on the design of a tunnel in the Swiss mountains, which was equipped with an elaborate ventilation system with smoke extraction and emergency exits to a parallel escape tunnel. According to a probabilistic analysis, based on statistical data and traffic prediction, a fire in that tunnel was to be expected on average once in a few years. However, after the opening of the tunnel, there occurred three fire incidents in the first three months, involving one burning truck, one bus and one personal car. Nobody was harmed, and the damage was very limited. The early detection and fire ventilation systems worked as foreseen and tested, tunnel users fled to the emergency exits, and the fire brigade was able to control and extinguish the fire quickly in all three cases. Who would have questioned the expensive safety measures? By the way, since then, no further fires have occurred.

---

[13] Compare with vertical axis of Fig. 15.

Fig. 14    Fire test in a Swiss tunnel, where three real fire incidents happened shortly after opening to traffic.

In Quantitative Risk Analysis, assumed values on probabilities and consequences have a huge impact on the calculated risk. The basic problem is that probabilities can only be usefully determined for events that happen in significant numbers. The lower the probability, the higher is the uncertainty about its value. The determination of small probabilities is actually mere guessing, since there is no statistically significant record. Even the most sophisticated mathematical models cannot take into account the occurrence of unforeseen incidents. When human behavior is involved, risks are even more unpredictable. Risks with an infinitesimal probability and extraordinary consequences are of highest concern to us, but in fact cannot be determined, only roughly be estimated. An exact calculation of incident probability would be possible where consequences result from multiple single events with particular probabilities according to known physical laws. Such is the case for instance for the failure analysis of technical facilities, where reliabilities of critical components, which are built in large series, are known from experience.

Just to complete the scheme: Low probability / low consequence risks should not bother us. High probability / high consequence risks don't exist for long, otherwise we would be eliminated soon.

## 5.5 How Safe Is Safe Enough?

The principal goal of risk analysis is to define a reasonable balance between safety and necessary costs to achieve the safety goals. The basic question is 'How Safe is Safe Enough'[14]. The society and its decision-makers must be willing to bear the residual risks exceeding predefined acceptance criteria. For instance, in Fig. 15 the result of a QRA for hazardous goods transports in road tunnels is displayed in a frequency-consequence-diagram, where the red lines define acceptance levels. Risks in the upper right part of the diagram are not accepted, and must be avoided by additional safety measures, whereas risks in the lower left part, below the red lines, are perceived as acceptable.

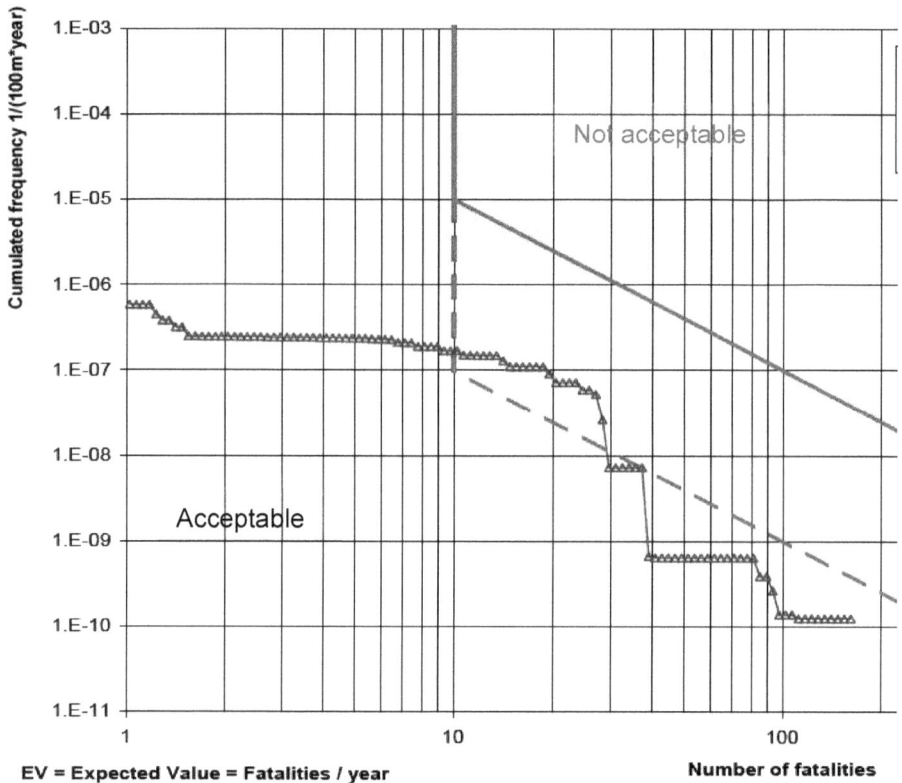

Fig. 15    Example of Frequency-/Consequence-Diagram from a QRA

---

[14] In fact, this would be a more appropriate title for this book, but is already in use in many articles and books by other authors (e.g. [41], [79])

With the number of fatalities per time as calculated by quantitative risk analysis for particular dangers and variants of safety measures, the efficiency or those investigated safety measures can be calculated as cost per life saved, and compared with a previously defined acceptance level. For instance we evaluated that the cost per life saved by additional emergency exits provided by a newly built parallel escape tunnel alongside an existing single tube road tunnel would exceed a Billion EUR per saved life[15]. On the other hand, simple regulations like mandatory motorcycle helmet laws may be much more cost-efficient, with an estimated cost of 2000 USD per saved life in the U.S. [131]. From my personal experience, I appreciate that very much, since once I had a motorcycle accident that I would hardly have survived if I hadn't worn a helmet, at least not without serious head injuries.

A descriptive example for the definition of a reasonable limit to safety measures are floodwalls on shores and riversides that are designed to withstand a specified flood level, based on sophisticated predictive models and statistical data. A 'once-in-a-hundred years flood' may define the design case, but there is no guarantee that the 'once-in-a-thousand years' flood will not occur just tomorrow, taking into account the increase of extreme weather conditions by climate change. For instance, in 1975 the worst disaster of structural failure ever occurred, when the Banqiao Reservoir Dam in China was breached. Built in the early 1950ies, the dam was designed to withstand a 'once-in-a-thousand years' flood. However, in August 1975 more than the mean annual amount of rain fell in only 24 hours in the aftermath of the third typhoon in that year, causing a 'once-in-a-two thousand years' flood. Over 700 million cubic meters of floodwater were released over the course of six hours. Approximately 26,000 people died from flooding and another 145,000 died during subsequent epidemics and famine. In total, 11 million residents were affected. Some reports estimate the number of fatalities to be as high as 230,000. Another vivid example of the application of a useful design case is the determination of the height of seawalls in Fukushima. In those examples the model was right, and the risk was known in principle, but not in its extent. The assumptions about decisive factors might be too optimistic in hindsight. However, as long as nothing happens, it is difficult to convince investors about the usefulness of abundant safety margins. The question 'How Safe Is Safe Enough' is answered differently before and after a disaster that exceeds the previously defined acceptance level.

---

[15] The mentioned risk analysis was not published. The road authorities ignored the results, since the escape tunnel was demanded by the guideline.

Some politicians claim a 'zero risk' policy. Unfortunately, we live in a world of limited resources, as most of us learn in early childhood. Therefore, in any endeavor, the available funds should be distributed efficiently. Resources invested in one particular measure lack for other applications, where they would possibly be more effective. In many countries, members from safety boards or emergency services, for instance the fire brigade, impose strict requirements on safety issues. I experienced many times that they would not accept any reduction in previously defined safety standards, and every human life must be saved, disregarding the necessary costs. Engineers, construction companies, equipment suppliers, and related services profit from such exaggerated demands. Everybody follows his or her own interests, and cheating is normal – to a certain extent [1]. As a ventilation engineer, I would profit from every tunnel being equipped with an expensive smoke extraction system. But can such an attitude be an incentive to do honest, satisfying work? Demanding excessive safety measures without considering costs and benefits must be questioned, taking into account that with money that is wasted for ineffective measures, many more human lives could be saved in other fields of risk mitigation, for instance disease prevention. That's why cost-benefit analysis should be applied in decisions with high impact.

On the other hand, which politician in a democracy can afford to openly admit that he is against safety measures, that would potentially save human lives? Even more when those measures are paid by an anonymous, remote entity. We rarely take into account that the state and its institutions represents our society, and is financed by all of us, at least those who pay taxes, road tolls, and other duties. In fact, demanding 'zero risk' or 'ultimate safety at any price' is anti-social. With Lewis' words [80]:

*'As a society we really can't afford to act as if life were truly priceless'*

## 5.6  Medical Check-Up

The following simple example shows how important the consideration of probabilities and underlying statistical data is[16]. Regular medical check-ups are
recommended to reduce your health risk. Particularly, early cancer diagnosis
might save many patients. Let's assume that on such a check-up, your doctor
tells you that you have been diagnosed with cancer type X. Since the test is
considered quite trustworthy, an operation would be necessary. Will you con-
sent?

A thorough analysis, based on the underlying probabilities, would be advisable.
The crucial question is, what is the base rate of cancer type X in the population? How frequently does it occur? For this example, lets assume that on average one in 100,000 people will get this type of cancer. How reliable is the
diagnosis? No testing procedure is perfect. All tests have a certain error rate,
which is usually determined by extensive investigation prior to approval. If the
test indicates cancer X, when you really have got it, then the result is 'true positive'. On the other hand, when you are healthy, and the test shows no signs of
cancer X, then it is 'true negative'. Those are the desired results. But then, the
result may be 'false positive', when it indicates erroneously the existence of
cancer X. On the other hand, when it does not show any signs of cancer even
if you have it, then the test is 'false negative'. For this example, we assume that
the reliability of the test is 99.9%, meaning that only 0.1% of the diagnoses are
supposed to be faulty, both positive and negative. With these numbers, a simple scheme can be set up for a total population of 100 Million people[17], displayed in Fig.  16.

|  | Have cancer X | No cancer X |
|---|---|---|
| Total 100 Mio. | 1,000 | 99,999,000 |
| Test positive | 999 | 99,999 |
| Test negative | 1 | 99,899,001 |

Fig.  16   Scheme of positive and negative test results

---

[16] This example is taught in any beginner's course in statistics, but unfortunately seems
not to be common knowledge among many doctors and patients.
[17] Admittedly a very large number, but useful to avoid fractions.

According to these assumptions, the probability that you really have got cancer X, when it has been diagnosed, is only 1%. According to Fig. 16, from 101 positively tested people, only one really has cancer, while the diagnosis for the other 100 would be wrong. Take into account that any measure that is necessary to remove the cancer will significantly deteriorate your life. Chemical or radiotherapy inflict a lot of damage to your health, since these treatments don't kill exclusively the cancer cells. Operations impose an additional risk. Doctors can commit medical malpractice and infections occur even in a modern hospital. What are the incentives of your doctor, considering that cancer treatment is a business with high profit rates? On the other hand, have you ever experienced the suffering of a beloved one in the last stages of cancer? How would you decide?

In practice, the probabilities, basic rates of diseases and error rates of medical tests are not always well known, and mostly only assumptions based on incomplete statistical data. Nevertheless, the magnitude as used in this example is realistic. Usually, in case of a positive result in the first test, a second independent test with a different method has to be applied to increase reliability. Hopefully you will never have to make such a decision.

## 5.7   Time Scales

*'There are times when the utmost daring is the height of wisdom'*
(Carl von Clausewitz, German military theorist, 1780 – 1831)

Beside probability and magnitude, another important aspect of risk is the time-scale. Life is a dynamic process. Risk is not a static value, but changes with time in many aspects. Changing circumstances make optimal solutions for a specific, outdated risk obsolete, and may even lead to new hazards. In the real world, whether in nature or in the man-made environment, everything has a limited lifetime. Buildings, systems and equipment are subject to deterioration, and must be maintained and refurbished with a permanent effort. Failure of components is often a stochastic process, and without maintenance and regular replacement of aging parts, any system will finally break down.

One of the first questions in any risk assessment is how imminent a possible threat may be. Being exposed to immediate danger, there is usually no time for thorough analysis, and decisions need to be taken urgently in the face of incomplete, dubious, and often completely erroneous information and high levels of fear, doubt, and excitement. In the midst of a disaster, it's too late to learn.

Identifying possible scenarios and developing, learning and training the right reactions should be strived for in advance. For emergency services and professionals in risky jobs, following automatized procedures by conditioning and extensive training is essential. Such procedures must be based on previous thorough analysis. However, reality is always different from theory and training scenarios. Situational risk awareness, flexibility and the ability to improvise and react to the unexpected are of utmost importance. Unfortunately, drills and training based on strict predefined rules are often contradictory to such attributes.

On the other hand, when there is time for careful analysis and possible countermeasures, it should be worked out without pressure. But then, risks are often underestimated or completely ignored, until the consequences become unavoidable. Natural risk perception is about a fast reaction to imminent dangers, not about abstract long-term threats.

The results of a risk analysis are usually referred to a defined time period, for instance the number of casualties per year (due to a certain risk). For most risks, exposure time is limited, therefore the overall risk is usually a mean value over a certain time, but momentary risk may vary according to actual exposure. Therefore, a simple and most effective measure for risk reduction is to limit the exposure time to the risk situation. For technical systems, the exposure time is usually given by the operational time of the device. Subsequent risks have to be considered after the end of the operation or the lifetime of the particular facility, for instance by dismantling procedures and the waste that has been left. Time framing is essential for risk determination. For instance, tunnels improve the safety of traffic infrastructure. When safety requirements lead to the delay of project realization, their effect might be contradictory, and the safety requirements in fact increase the risk in the long run. If there is a slight chance that something might happen, then it will happen sooner or later, when the time frame is set long enough[18]. A time delay between an event and the manifestation of its effects may lead to neglect or significant underestimation of risks, because the originators of that event will often not be affected. It's impossible to fully understand all aspects of interactions and risks in complex systems with long term effects. Slow, continuous, invisible processes, causing damage in the long run, are regularly overlooked. That is the case for most health issues or ecological systems. Such systems are mostly non-linear, with many unknown parameters to be taken into account.

---

[18] Zero probability in an eternal timeframe is an issue of theoretical mathematics.

Lewis proposes to discount future risks in a simple manner, taking into account that the delay between cause and effect is a crucial factor [80]. Some hazards lead to immediate risks, for instance traffic accidents. If you drive too fast, with little distance to other vehicles and obstacles, you might not be able to react fast enough. Other risks imply significant delay, which may lead to negligence. When you smoke today, you might get lung cancer in 20 years. Many risks that cause possible damage in the long run affect our children and grandchildren, but most probably not today's decision makers.

Then there are risks with time scales far beyond human imagination, for instance resulting from nuclear waste disposal. Since radioactive contamination remains a threat for tens of thousands of years, there might be moral concerns about whether we should leave such a problem to our heirs. On the other hand, deposits in geologically stable formations are quite safe in the long run, and why shouldn't we suppose that later generations in the far future won't be able to solve any related problems, should they occur? Our ancestors obviously did not have any such concerns and left some massive ecological problems to us. Humans have contributed to the extinction of many species of plants and animals that are lost forever by now. In prehistoric times, human hunters were probably one of the main causes that led to the demise of the megafauna. Huge landscapes were deforested in order to cultivate the land for agriculture, with subsequent desertification after depletion of fertility and erosion. Wood was the primary source of energy and construction material. Only thanks to fossil fuels, both the industrial revolution and sustainable reforestation were possible. What will come after the fossil fuels run out?

## 5.8   Limits of Risk Analysis

The full impact and consequences of any human endeavor are never known in advance. As described, not all relevant aspects of safety are covered by common risk analysis methods, and knowledge is mostly limited. The worst disasters may be caused by single incidents whose probability is considered to be negligible (but not zero), or by events that are not foreseeable at all. Risk analysis methods taking into account low probabilities and uncertainties are based on sophisticated mathematical models [6], [10]. However, the more complex and accurate a risk analysis model is, the more it is dependent on assumptions on modeling, parameters and boundary conditions, therefore its error-proneness increases.

To evaluate the effect of variation of probabilities, initial and boundary conditions on the calculated risk, a sensitivity analysis with varying parameters should be worked out. But that costs time and money, which is often not reimbursed.

Comprehensible results from Quantitative Risk Analysis are useful as a basis for decision making, using a unified, clearly defined criterion, while taking into account its limits. Unfortunately such results are rarely understood nor accepted by executive decision makers and the public.

| Type of Risk | Applicability of QRA | Example |
|---|---|---|
| Unexpected | Not covered | 9/11 Attack |
| Expected but Unknown | Not covered | Exposure to low concentrations of carcinogens |
| Low (unknown) probability High consequence | Based on assumptions → Estimation by QRA (magnitude) | Technological incidents Natural disasters |
| Regular hazards Known probability Known consequences | Based on experience /statistical data → Predictions by QRA (in average) | Traffic accidents |

Fig. 17    Applicability of Quantitative Risk Analysis (QRA)

Risk analysis is based on the assumption that the process leading to possible incidents is understood and can be modeled. In contrast, when the model is wrong, systemic risks lead to unexpected consequences. Such are not covered by risk analysis in principle. An example of such a systemic risk is the worldwide financial crisis in 2007 and the following years, taking into account that the financial institutes employed legions of well-paid professional risk analysts who obviously did not really understand the issue.

Risk managers using quantitative methods often miss the forest for the trees. Risk management is about the forest. Quantitative methods should not replace logical analysis. Furthermore, decisions on risk issues cannot be made based exclusively on rational analysis when it is contradictory to people's feelings, as shown in the Ford Pinto Cause.

By the introduction of additional factors for events with low probability and high consequence, human feelings are taken into account in the results of risk analysis. Such are designated for instance as 'dread factor' or 'risk aversion factor', being rather a subject of political discussions than an objective and scientific definition. Often, the decision has been made beforehand, based on subjective arguments, and the risk analysis serves as a mere token gesture to justify the desired results.

## 5.9    Ford Pinto Cause

The Ford Pinto was a car built in the U.S. between 1971 and 1980. Due to a design flaw, the fuel tank could be easily damaged in the event of a rear-end collision, eventually leading to a fire. 27 people died as a result of such fires in 10 years, and many more were injured. Taking into account that in the same period, almost half a million people lost their life on U.S. roads, the 27 fatalities (representing 0.005% of the total body count) would be negligible.

Fig. 18     Ford Pinto during a rear-end crash test (*Public Domain*)

The engineers at Ford proposed modifications that would improve fire safety, but after performing a cost-benefit analysis, the management decided that the improvement would be too expensive. The cost to redesign and rework the Pinto was estimated at 137 Million USD, while the value of a human life was determined to be approximately 200,000 USD and a serious burn injury approximately 67,000 USD. Over 2 million Pinto's were built. If the 27 dead could be saved by the proposed improvement program, the cost per saved life would be approx. 5 Million USD. From the point of view of production costs, the decision not to fix the design flaw based on a technical analysis can be justified, but the affected people, the media and lawyers saw things from a different point of view. Some of the victims sued Ford, which resulted in significant compensation payments. Discounting human lives in order to save production costs was not acceptable to the public.

# 6    Risk Mitigation

## 6.1    Basic Considerations

When a recognized risk is considered not to be acceptable, it can be reduced by appropriate measures. Measures can be implemented in the behavior and appropriate training of individuals, in organizational provisions in groups and societies, and in the application of technologies. Safety measures are usually applied for risks that have been clearly identified and analyzed. Risk can be reduced, but never completely eliminated. You can control only something that is under your control!

After each disaster, the question for causes arises, and how it could have been prevented. Were there any appropriate safety measures to prevent it? Were they not sufficient, or not applied correctly? Possibly it was simply bad luck. The importance of mere chance is often underestimated in risk issues.

First of all, the exposure to a possible threat should be reduced or even avoided. For risks that are caused by human behavior and technology, probabilities can be minimized by primary measures, referred to as prevention, and consequences (after the incident has happened) can be reduced by appropriate secondary measures. In health issues, primary measures are called prophylaxis and secondary measures therapy. Insurances, reimbursing at least partially the damage, can be seen as tertiary measures. But for all risk mitigating measures, a price has to be paid. For risks with relatively high probabilities, with a statistically relevant data basis, the effect of risk mitigating measures can be proven, as for instance road traffic safety. The effect of measures can also be easily demonstrated for simple technical systems, but with increasing complexity, any measures may lead to unforeseen consequences.

Safety is achieved by permanent effort of awareness and active risk mitigation. But only one single event, whether caused by negligence, by an unforeseen incident or by deliberate sabotage, can lead to a disaster. By that, all previous efforts become finally useless. After any disaster, the related safety measures are discussed and analyzed. New improved safety measures are developed, probably for an event that would never happen again. Does it never happen again due to the safety measures or because it had simply been a unique event?

Effective risk mitigation measures prevent incidents, but something that does not happen is a poor argument in an environment where financial pressure and tight time schedules prevail. The efficiency of risk mitigation measures that prevent single incidents from happening is impossible to prove. Quantitative risk analysis may be used as a useful tool to compare the impact of different safety measures on a system by modeling and simulating various, hopefully significant scenarios. However, the absolute efficiency of a safety measure can never be proven; it may only be compared to a fictive situation without the measure when all other boundary conditions are equal, which in practice is hardly ever possible. Ideally, the most cost-efficient solution to achieve the (previously defined) acceptable safety level shall be chosen. In reality, safety measures are often chosen by personal preferences of decision makers, strongly influenced by momentary media coverage and commercial interests of providers.

Another aspect is that many risk-mitigating measures protect the infrastructure, but have little efficiency on life saving. Some even imply additional risks to people. Avoiding property damage is sometimes given a higher importance than protecting human life and health[19], not only by the property owners but also by the endangered people. This can be observed for instance during fires in road tunnels, when drivers are reluctant to leave their vehicles. In the Gotthard road tunnel fire incident many victims were found dead in their cars, despite having had the opportunity to escape to adjacent emergency exits.

## 6.2 Primary Measures

The goal of primary measures is to prevent any incident that would lead to damage from happening, or at least reduce the probability of occurrence. The first and most important measure is to avoid or reduce exposure to risks, simple awareness of possible risks and appropriate alertness. But that should not lead to a state of paranoia.

Many primary measures are about influencing human behavior. Not giving people the opportunity to inflict damage is the best way. In case of unhealthy food and drugs, avoid them or limit the dosage. Any individual can reduce his health risks by not eating sweets and smoking cigarettes. The risk of bad people doing deliberate harm to you can be limited by avoiding their company, or by strict access control. That's why you lock up your front door.

---

[19] Of course that is mostly denied in public communication

Any situation in which bad people have excessive power should be avoided. Unfortunately, for citizens in brutal dictatorships, subordinates of a ruthless company boss, and subjects in violent relationships, this advice is difficult to follow in practice. Beside hostile fellow humans, contagious diseases were among the worst hazards during the whole history of mankind. Primary measures against that risk are hygiene, a functioning system for disposal of waste and sewage along with vaccination and medical treatment by antibiotics.

As far as the example of road traffic safety is concerned, the most effective primary measure would be to use the train instead of the car, or to drive slower. Preventing incidents can often be achieved by taking away possibly dangerous elements. Highways are made safer by avoiding crossroads or sharp curves. The safety of mountain roads exposed to rockfall can be increased by building tunnels and galleries, or at least by closing the roads during critical seasons. Warning signs, traffic management systems, legal speed limits and signaling on roads lead drivers to coordinated behavior, reducing the probabilities of congestions and accidents, but take effect only when respected by the drivers.

## 6.3    Secondary Measures

Secondary measures reduce the extent of damage, after an incident has happened. They do not prevent the incident from happening, but are required when the primary measures have failed. Therefore, their efficiency in risk reduction is principally lower than that of primary measures. Typical examples of secondary measures are early warning systems against natural disasters, fire safety systems like smoke detectors, sprinklers, smoke exhaust ventilation, emergency exits, airbags in cars and ejection seats in airplanes, or medicines in health issues. Wearing a helmet, training self-defense or having a first aid kit readily available can also be seen as secondary measures for risk mitigation.

By deployment of emergency services like the fire brigade, the ambulance and law enforcement, the damage in case of incidents can be limited. When primary measures work, secondary measures are only rarely, if ever, required to be applied. But how can you prove that they work? They should be regularly tested (e.g. safety equipment) or trained (e.g. emergency service personnel) as realistically as possible, preferably under stress conditions, but without causing undue damage by the test itself.

In a rational decision making process, primary measures should be given a higher priority than secondary measures. In reality, people mostly focus on secondary measures. They give you a good feeling by their mere existence, but you hope you will never need them. Sometimes their value is rather imaginary than real. Safety measures in the form of technical provisions can be passive like helmets, emergency exits and fire protection isolation, or active like traffic management systems, ABS and sprinklers.

The benefit of passive systems is that if applied correctly, usually nothing can go wrong with them, and only deterioration with the passing of time has to be taken into account. On the other hand, passive systems are fixed and cannot be adapted to changing situations. Active systems require mostly energy supply, regular testing and maintenance to be functional. In active systems, possible failure is intrinsic, according to Murphy's law.

## 6.4   Drawbacks

The most obvious drawback of safety measures is that they cost time and funds, without direct return of investment, compromising the primary goal of any endeavor or facility. Apart from that, any safety measure may increase risk in aspects that were neglected or not considered at all, for instance unwanted side effects, erroneous operation or failure.

Some safety equipment may even increase the risk by the possibility of malfunction. Such weak points should be discovered and addressed by a thorough analysis. There are always situations in which safety measures have benefits, while in other situations they could cause damage. For instance, airbags in cars significantly reduce harm to people in case of collisions at medium speed. At low speed, a simple safety belt is sufficient, and at high speed, the consequences of collisions are severe anyway. However, in case of unwanted release while driving, an airbag can cause serious accidents. Therefore, reliability of the activation mechanism is of utmost importance.

A similar case are sprinkler and fog systems in tunnels for fire suppression. The consequence of erroneous malfunction of such fixed fire fighting systems is most important. Sudden release of water spray in a tunnel under traffic, frightening drivers, can result in serious accidents. The probability of such a malfunction is quite low, assuming that the system is reliable and well maintained. But the probability of a fire incident in a specific tunnel, when the operation of the fixed fire fighting system would be useful, is also low. Both the fire and the release of sprinklers under traffic lead in most cases only to little damage, but both can also cause disasters with many fatalities.

In fact, any quantitative risk calculations regarding sprinklers in road tunnels would be mere estimates with unknown probabilities of unknown consequences. What really counts is that sprinklers can limit material damage, but also lead to significant costs, in particular when taking into account the necessary maintenance and testing during the whole lifetime. The decision is about money, not about people's safety, and the usefulness of fixed fire fighting systems is limited to very special cases where a collapse would lead to unacceptable damage, for instance urban tunnels under buildings.

My personal favorite example of safety equipment that is not only unnecessary, but in fact may increase the risk, is smoke extraction in tunnels with unobstructed unidirectional traffic. The arguments are described in chapter 11.6 and more in detail in the Tunnel Ventilation Compendium [109]. Operators tend to use equipment only because it is there, without questioning and understanding what it is good for. Erroneous operation of a smoke extraction system may lead to tunnel users, who are blocked in front of a fire incident, to be exposed to toxic smoke. Without extraction, relying on the longitudinal airflow, they would stay safe. Such situations happen regularly, not only by fault of a stressed operator in an emergency situation. Unsuitable smoke extraction systems are demanded by many national safety standards, and I have observed that even experienced ventilation designers prescribed operational conditions that would lead to a disaster.

In any risk assessment, failures of safety systems have to be taken into account. Risk reduction by prevention, provision of temporary replacement systems, increased readiness of operation staff and emergency services, and provision of redundant systems must be considered. Unforeseeable failures during operational requirement in case of an incident can in principle be compensated only through system redundancy. Irrational human behavior is often neglected in the assessment of safety measures. The psychological effect of risk compensation when people relying on safety measures take higher risks may also be seen as a drawback of safety measures. The transfer of responsibility from a person to a technical measure, which is again designed and implemented by other humans, is not always the best way to go.

Moreover, the usefulness of risk mitigation measures strongly depends on the circumstances. Mitigating one possible threat may lead to other risks. For instance, reinforced security doors, preventing any attempt to enter the cockpit from outside when locked, would have prevented 9/11. As a consequence after that terrorist attack, such security doors subsequently have been required in all commercial airplanes.

Unfortunately, such doors facilitate pilot suicides, for instance on the 24<sup>th</sup> March 2015, when the co-pilot of Germanwings flight 9525 locked the door while being alone in the cockpit and crashed deliberately into a mountain ridge in the French Alps, killing all 150 people on board.

## 6.5   Insurance

Instead of being mitigated, the risk can be transferred to another party, at least to some extent, in form of a risk-pooling arrangement. Such a party can be a social group (for instance your family), a commercial insurance company or the state. The principle is the same: The costs for the consequences are distributed between many participants and by that diluted. The insurance reimburses the damage, at least partially, in case of an incident, by collecting funds from the many parties that don't suffer any damages.

Insurance is similar to a lottery, where the money is collected from many participants, who hope to win the jackpot. But only one or a few can win. For insurance, the prize is the compensation in case of damage. In lottery, the probability to win is mathematically clearly defined and easy to calculate, but insurances operate within the uncertainties of real life. In simple words, the insurance premium for each insured is the expected monetary risk, divided through the number of insured, plus an additional margin for the unexpected and the profit. Obviously there are mostly profits in abundance for the insurance companies, which base their business model on irrational human risk perception, particularly the exaggeration of small risks. Nevertheless, insurance companies can go bankrupt by one or multiple incidents, if the damage is large enough. Therefore, reimbursement is usually limited. When there is no insurance available, you should be very cautious.

Being insured may lead to a decline in personal responsibility. Knowing that a possible damage will be reimbursed reduces incentives to be prudent and may encourage people to take higher risks. Therefore, insuring bodies often require obligatory risk mitigation measures by contractual terms, thus increasing safety by commercial interests, but also limiting the range of self-determination of individuals. In this respect, insurance companies have a strong economic power in our society, particularly on issues where insurance is mandatory, for instance car owner's liability, homeowner's fire insurance, or obligatory health insurance as in most European countries.

## 6.6    Confidence

Confidence is a good feeling, but not always justified. Whether safety measures really work and fulfill their purpose is not a matter of course. When safety systems are never really used, but become difficult to maintain, unrelia- ble, or do not function properly, operators tend to bypass them or even remove them completely. Only what has been proven in real incidents and in regular stringent tests, ideally under different stress scenarios with varying boundary conditions, is supposed to work reliably. Fortunately, real incidents, in which safety equipment is required, usually don't happen very often. If they do, it is under momentary random boundary conditions. For risk mitigating measures that have passed only a few tests (or none at all), or many tests but under simi- lar conditions, the temporary absence of failures might be pure coincidence. There always remains a principal uncertainty due to the problem of induction, neglecting the unexpected. An accident not happening is no guarantee that it won't happen. Every situation is unique, and there is no guarantee that the next situation will be the same. Heraclitus, a Greek philosopher (535 – 475 BCE) wrote:

'*You cannot step twice into the same river*'

Large numbers must not be mistaken with a long time: A concept or system which is inapt might last for a long time, as long at it is never really required. Success breeds complacency, which is the first step to failure. That's the case for many types of safety equipment, which is built for extraordinary events that never happen – but might happen one day. Defects in equipment that is practi- cally never in operation become first visible when an incident occurs, but then a failure can have disastrous consequences. That's why testing is so im- portant. Defects that occur while testing or in real incidents are usually kept secret, and only the worst mishaps and disasters are made public. In contrast, for regularly used technologies and equipment, any faults are revealed soon, and a steady progress is guaranteed when the right lessons are learnt from failures.

Personal and material damage can be reduced by appropriate reaction of the people involved. Previous information and preferably regular training to gain experience are essential to perform well in hazardous situations. Like in tech- nical systems, practice under stress conditions gives people the confidence that they get a better chance to deal with dangerous situations. This is espe- cially important for members of professional emergency services like firefight- ers, policemen and emergency doctors.

Personal abilities, acquired by appropriate learning, extensive training and practical application, are usually more important than equipment. In this respect, education and training of emergency services, operators and drivers would also be a most important safety issue, which is strongly neglected. After the fire in the Mont Blanc road tunnel in 1999, billions have been invested in tunnel safety equipment worldwide, but have you ever practiced evacuation from a tunnel filled with smoke?

In medieval times, confidence was achieved by sale of indulgences, providing a good business to the omniscient Catholic Church. It promised to mitigate the risk of your soul going to hell as a consequence of your sins, which was feared as one of the worst hazards by then. It was more comfortable to pay for indulgences than to make an effort to follow the principles of a righteous and faithful life. Modern sale of indulgences is not very different from that. Our society, namely some executives who are in charge of public funds, is willing to invest enormous sums into the mitigation of palpable, but low risks, particularly those that have recently been covered by the media, because it gives them confidence. The same decision makers often don't like paying even a little tip for independent, thorough analysis and useful risk management. Even worse, funds for testing and training are limited. Objective risk analysis might conclude that some of the 'safety measures' are of little use and often don't work, and on the other hand important risks are neglected.

An important issue is to try whether the implemented risk mitigation measures really work under all circumstances, and to apply possible improvements. Quality management standards prescribe a continuous process of assessment and improvement, but in practice are often only a token gesture. Having invested time and money into a concept or system that promises safety gives you confidence. Analyzing, questioning and testing whether it really works might let you doubt. That's why people are reluctant to prove whether safety measures really work. In fact, many don't!

On the other hand, as shows the Chernobyl nuclear disaster, testing of safety measures can also terribly go wrong.

## 6.7 Chernobyl Nuclear Disaster

Perrow, an US sociologist (*1925) examined the 1979 Three Mile Island nuclear accident in the U.S. where a disaster could be barely avoided after a partial nuclear meltdown. In the book that he published subsequently [97], he wrote in 1984: '... *the probability of a nuclear meltdown with dispersion of radioactive materials to the atmosphere is not one chance in a million years, but more like one chance in the next decade.*' Unfortunately, he was proven right only two years later.

On April 26, 1986, reactor no. 4 in the nuclear power plant of Chernobyl in the Soviet Union (today on Ukrainian territory) was destroyed by a steam explosion and subsequent fires that lasted for about 10 days [57]. At least 5 % of the radioactive reactor core was released into the atmosphere and downwind, spreading over a large part of Europe. The main causes were a faulty reactor design, inadequately informed and trained personnel, and a chain reaction of mistakes made by the operators and things going wrong during a test, finally leading to an uncontrollable power surge.

Fig. 19    Destroyed reactor at Chernobyl Nuclear Power Plant *(Public Domain)*

Two plant workers died on the night of the accident, and a further 28 people, mostly fire fighters, died within a few weeks as a result of radiation poisoning. Subsequently, there were thousands of casualties due to long-time effects of radiation syndrome, mainly among the personnel doing the cleanup operations (the 'liquidators'). The inhabitants of surrounding villages and towns were evacuated only after a significant delay of 36 hours after the explosion of the reactor. Many of them later suffered from cancer. The economical conse-quences of extensive radioactive contamination in many European countries can hardly be estimated. The Soviet regime has never published official data. Any numbers of casualties in actual publications are mere estimates, depend-ing on the intent of the authors.

Chernobyl is recognized as the worst nuclear disaster in history[20]. Human er-rors played an essential part. The plant operators were about to simulate an electrical power failure, consciously disabling important protection equipment. They did not comply with operational procedures, obviously not being aware that they would bring the reactor into an uncontrollable condition. Such behav-ior had been encouraged by previous experience. However, the operators are not to be blamed alone.

Anatoly Stepanovich Dyatlov (1931 – 1995), the former Chernobyl deputy chief engineer and supervisor of the fatal experiment, has pointed out that the reac-tor didn't conform to its design norms [38]. When a nuclear reactor can be brought to an explosion under operating conditions, the design has obviously not been well thought out. A disastrous accident would have happened sooner or later anyway. The key issue were the protection rods that should have served for the emergency shutdown of the nuclear reactor. In Chernobyl, those rods even accelerated the nuclear reaction in the first phase of their insertion, and when released in the last attempt to avoid the looming catastrophe, they got stuck in that first phase. With Dyatlovs words:

*'The reactor was blown up by the emergency protection'*

---

[20] Probably there had been even a worse incident on the 29th September 1957 near Kyshtym, where up to 100 tons of high-level radioactive waste were released, but the Soviet dictatorship kept it secret and until today there is little reliable information.

## 6.8 Diversity and Decentralization

Even with the most elaborate risk mitigation measures, failures and incidents will happen inevitably. Therefore, the resulting damage should be limited in principle. Limiting size means limiting effects, in a positive and a negative way. A conglomeration of small entities, where failure or destruction of one or a few single components can be accepted, because it does not lead to complete system failure, is more resilient against hazards than a big, centralized unit. Diversity and decentralization are some of nature's most important principles, and may be applied to different types like ecosystems, technical facilities, investment portfolios or human organizations. In principle, diversity means providing a lot of reserves and flexibility. If one entity fails, another one will take its place.

Diversity and decentralization are also important in assigning competence for decision making, accompanied by responsibility, in human organizations. Strictly hierarchical organizations tend to break down completely, when a single decision maker fails. Companies with encouragement of leadership and assignment of responsibility and appropriate knowledge to competent employees mostly prevail in a competitive environment. The same is true for military organizations. One of the reasons why Germany became a successful military power in the 19th and 20th century is the concept of mission-type tactics, as developed by Helmuth Karl Bernhard Graf von Moltke (German General and Strategist, 1800 – 1891). Rather then having to strictly follow detailed orders, commanding and noncommissioned officers are given goals that have to be achieved. The accomplishment of those goals is left to their initiative according to their individual assessment of the situation. Subordinates on each level are given as much decision making independence and subsequent responsibility as possible, within the limits of the senior commander's intention.

On the other hand, in many traditional societies with strict hierarchies, the masses of common soldiers are deliberately left without education and have strictly to follow orders, which must not be questioned, given by an elite of officers. Such armies lose wars due to their inflexibility and lack of motivation, as is described for instance by De Atkine [28][21].

---

[21] Unless overwhelming their opponents by sheer mass, sacrificing a high toll of human lives as cannon fodder.

Other examples of diversified, decentralized systems are the competitive free market of supply and demand, which is one of the very sources of progress and wealth, and the internet, which enabled an unprecedented availability of information, based on a decentralized distribution of storage and communication lines. Both systems are not perfect and have many flaws, but in principle work well – as long as the government, representing the society, enforces strict rules when safety and common goods are at stake.

Multiple small entities are not as powerful as big, centralized organizations, but failure of an entity does not affect the system as a whole, therefore the possible consequences are limited. Reserves, self-sufficiency and complex decentralized organizational relations bind funds in terms of time and money. Any manager looking for short-term efficiency and maximization of profits will reduce reserves and focus on the core business. Diversity is contradictory to that and therefore has a hard stand against economic pressure. It must be pointed out that in centralized systems, a high level of efficiency and individual safety can be achieved, by reducing the probabilities of failures. But unfortunately, when failures happen, they may lead to disastrous consequences for the whole system, even when the probability is much lower. In contrast, in diversified systems, the individual risk of failure is higher, but the common risk of the whole organization is lower. Others succeed and can support you when you failed. This is the basic principle of general welfare based on a social market economy. In contrast, all centrally planned economies have failed sooner or later, which was one of the basic reasons for the collapse of the Communist dictatorships in Eastern Europe and the Soviet Union in 1989 and the following years. History has shown that in the long run, diversified systems prevail. In this respect, the loss of cultural and economical diversity by globalization may impose a serious, unknown risk.

Decentralization and diversity are key success factors of Switzerland's political stability and prosperity. Democratic decision making processes are worked out locally on the level of communities and Cantons that are joined together in a Federation. It's hardly comprehensible that in the past years, this diversity is given up for the sake of better efficiency and centralization – at least at first glance. A good example from my professional environment is the centralization of highway projects following the new financial equalization between the Cantons. Before that, road projects were worked out by local authorities, each one with different standards and project management procedures. The project managers in local administrations shared their offices with the operators and maintenance personal, and were close to the projects they were working on.

Some project teams were formed by competent, experienced engineers pursuing good technical solutions, others showed incompetence and poor performance. Even then, federal standards had to be followed, but the implementation in practice was up to the local project team. But centralization in the course of new organizational structures made things worse. The consequence was an increase in bureaucracy, formalism, and costs – contradictory to what had been promised to the tax paying voters. A completely new organization was built up, mostly by recruiting new civil servants without appropriate experience. Only few members of the previous local professional teams have been employed by the new federal bodies. Technical competence has significantly relapsed, and time and money are wasted for investigations about previously well-known issues. A process of overregulation has been accompanied by an increasing absurd red tape, developing into an uncontrollable mess. Many projects were delayed and costs rose massively. However, for some big engineering companies it seems to be good business to provide administration services to this excessive bureaucracy.

## 6.9    Automatisation

Reducing the risk of human error can be achieved by transferring tasks to an automatic system, which is supposed to be safer than human operators. For instance in cars, assistance systems like EPS and ABS take over when the driver's reaction would not be appropriate. In production and power plants, computerized, automatic control systems have mostly replaced manned control centers. The control software can achieve an enormous complexity. The software, meaning the manpower invested in evaluating, programming and testing predefined states of operation, is usually more expensive than the hardware. The effort to find and eliminate all hidden faults, which are intrinsic in any software, may rise far beyond feasibility.

Automatic systems are designed and realized only by humans and are therefore prone to error too. The limitless number of unknown facts, hidden faults and alternative histories cannot be covered by predefined procedures. The unforeseen might happen as well. The probability of errors can be minimized by thorough analysis and a continuous improvement after testing and operation under stress conditions. However, in case of unforeseen events beyond the predefined scenarios, human operators should either take control, or the system will act on its own, hopefully getting into a safe and stable state, but possibly into a disaster.

For manual operation, which overrides the automatic, system comprehension and the ability to react quickly and in the right way would be essential. Unfortunately, those abilities are neglected when relying too much on automatisation or following standard procedures. Appropriate instruction and training of operators would be necessary, but is often omitted to save time and money.

For instance, civil aviation has been a success story in significantly reducing risks from human errors by applying automatic systems. By today, commercial airplanes have reached an extraordinary high level of operational safety. All relevant systems for operation and flying of modern aircraft are automated, and the pilots are in fact only needed in extraordinary cases. In this respect, modern pilots are sophisticated system operators, but they have lost the abilities and experience to manually fly an airplane, when it would be required. Without experience from regular training, humans can behave extremely irrationally in stress situations. In the case of Air France flight AF 447, the incapacity of panicking pilots cost 228 human lives.

## 6.10   Pilot error

Flow stalling is an aerodynamic effect which can occur for instance on the blades of ventilation fans, as well as on the wings of an aircraft. Mostly, it is unwanted, and can have catastrophic consequences. When I was young, I got some experience in flying paragliders. When a paraglider gets into stall, its canopy collapses due to lack of pressure, and you enter into a free fall. This can be applied deliberately to lose height quickly, as might be necessary for instance in case of an approaching thunderstorm. I tried out to stall while still in basic training, which was not to the pleasure of my instructor. Free falling is a thrilling experience. When you release the control handles, the canopy will open automatically, so that the glider will get into a stable flight by itself. Of course, for such a maneuver, you need sufficient height over ground. The situation is similar with most airplanes: By getting a stalled plane into a swoop, it gains impetus and finally ends up in a stable flight path. This is practiced by hobby and sport pilots occasionally, but how often is it trained by commercial pilots? Can the structures of a giant modern aircraft even stand such a straining maneuver?

Once I got into a stall with my paraglider quite unexpectedly due to heavy turbulences after the start from a mountain ridge. Half of the canopy collapsed immediately and I lost control, falling down without any chance to recover quickly. Luckily, I flew only a few meters above a smooth, steep slope that was covered with snow.

On the impact, the free fall just passed to a fast slide, which was soon slowed down by the snow. I didn't suffer any serious injuries. From my point of view, I had taken a limited risk, like in the avalanche incident described in chapter 8.1. I would not have dared to start over rocky ground under such wind conditions[22]. However, my little adventures are not the topic of this chapter, but another, rather tragic accident, which shows the limits of automatisation and the effect of incomprehensible human error, as described in the report of the French Civil Aviation Safety Investigation Authority [8].

Automated flight-control functions removed a great deal of uncertainty and danger from aviation. For instance, the modern Airbus series passenger aircraft are perceived as one of the safest and most sophisticated types of commercial aircraft in the world, equipped with state of the art automatic flight control systems, including a digital fly-by-wire (FBW) control system, and Sidestick control like in modern fighter jets. Further automatic systems integrated in the Airbus planes are the Electronic Flight Instrument System (EFIS), which covers navigation and flight displays, the Electronic Centralized Aircraft Monitor (ECAM) and a 'Flight Envelope Limit Protection System' which prevents maneuvers from exceeding the aircraft's aerodynamic and structural limits. During a complete loss of airspeed information, however, that system reverts to manual control, and the airplane behaves much like a conventional airliner, forcing the pilot to take over.

On the first of June 2009, on a flight scheduled from Rio de Janeiro in Brazil to the French capital Paris, Air France flight AF 447 flew into clouds associated with thunderstorms over the Atlantic Ocean at 35,000 feet. In this situation, all three speed sensors became iced over one after another, and subsequently the measurement failed[23]. Failure of airspeed information triggered the aircraft's autopilot and auto thrust systems to disengage. In this situation, if the pilots had done simply nothing, the plane would have followed its stable flight path. The two copilots were in the cockpit, while the captain rested outside. All three were experienced, highly trained professionals. But in fact, training to manually fly a large commercial airplane without relying on the information systems had obviously been neglected.

---

[22]Some of my former colleagues, ventilation engineers and hobby paraglider pilots, were not so lucky and occasionally suffered more serious accidents.

[23] By the way, the same effect of blocked airflow sensors regularly happens on fan and air duct monitoring systems too.

After loss of the speed information, one of the confused pilots pulled the side stick backwards and stayed in that position, increasing the plane's angle of attack to a degree where it got into a stall condition. This was the one single error that caused the following disaster. Subsequent alarms were released by the information system, but inexplicably, the pilot did not push the side stick back forward, but persisted on his hold. The other pilot tried to override his colleague's switch command. Panic, chaos and confusion between the pilots broke out in the cockpit. They seemed unable to comprehend the nature of the problems they had caused. Flying in a dark night over the ocean, there was no visual information available. The stall warning turned on and off as the aircraft fluctuated amid high angles of attack at very low horizontal airspeed levels. The crew failed to recognize that the aircraft had stalled and consequently did not give inputs that would have made it possible to recover from this danger-ous state. Free falling from a height of over 10 km above ground, the plane finally crashed on the water surface. Nobody had a chance to survive.

A simple but persistent mistake on the part of one of the pilots had caused the loss of a modern, fully operational aircraft without any serious adverse external influences.

Fig. 20    Recovery of a piece of debris from Air France flight AF447 *(Reuters)*

# 7 Plans, Rules and Laws

## 7.1 Planning

Planning means thinking ahead. The ability to do so is a basic characteristic that distinguishes us from most other animals. Imagining possible consequences of actions, based on an abstract model of the future and assumptions on underlying parameters, is the basic issue of human brainwork. That sounds complicated, but actually that's what we do – deliberately or unconsciously – most of the time. Unfortunately, reality may not evolve exactly according to our plan. Therefore, plans should be adjusted to reality in an ongoing process, taking into account new, formerly unknown aspects, and learning from errors, which are unavoidable. Most importantly, you have to consider that your plan may fail. Any plan must be flexible enough to be changed, adapted, or even completely given up during realization. Flexibility and the ability to improvise are essential. This has been recognized by brilliant thinkers and practitioners during the whole history. Lets quote them:


*'Malum est consilium, quod mutari non potest'*
*(It is a bad plan that can not be changed)*

Publilius Syrus (Roman writer, 85 – 43 BCE)


*'No plan of operations extends with any certainty beyond the first contact with the main hostile force'*

Helmuth Karl Bernhard Graf von Moltke (German General, 1800 – 1891)


*'Plans are useless but planning is indispensable'*

Dwight D. Eisenhower (U.S. General and President, 1890 – 1969)


in a simpler form:

*'Everybody has a plan until they get hit'*

Mike Tyson (former U.S. heavyweight boxing champion, born 1966)

The engineering principles described in chapter 12 are in fact part of the planning process, taking into account that things often go wrong.

This paragraph had been written while waiting at a small airport. I had been there on a business trip as technical supervisor for a long road tunnel project. After two days of meetings on the construction site, I was supposed to fly back on the same evening. Being exhausted due to endless discussions about the implementation of rather debatable design requirements, I was keen on getting home and getting some rest. Arriving at the airport, the flight was announced to be delayed. After waiting a few hours, it was finally cancelled. The airline provided an overnight stay in a hotel. The next morning, a taxi should have picked me up. It did not arrive, so after half an hour of waiting, being afraid that I would miss my flight, I ordered another taxi. Arriving in a hurry at the airport, I was informed that the morning flight was delayed too. After another hour of waiting, the airline informed me that I would miss my connecting flight. Finally they booked a connection on another route, and after a day lost at airports and on airplanes, I finally arrived in my office. Travelling time by car would have been shorter this time – but at a higher accident risk. Things had gone wrong, but it could have been worse.

## 7.2    Rules, Standards and Laws

*'Talent and genius operate outside the rules, and theory conflicts with practice'*

(Carl von Clausewitz)

Laws protect the members of societies from greed, ruthlessness and short-sightedness of single particulars or groups. A functioning legal framework, enforced by an independent execution agency, is one of the most important basics for a functioning society. The alternative is anarchy, as in many societies, where every dispute can lead to the outbreak of violence.

The community is responsible for the definition and, most importantly, consequent enforcement of safety goals. This community may be a state, a non-governmental organization, or a business company. Technical rules and standards are worked out as a means of fast, standardized decision making. Such rules have a huge impact on the choice of goals, technical solutions, concepts and costs. Without technical safety standards, economic pressure would lead to poor quality and deliberate suppression of hidden risks – what it actually does. Therefore, the elaboration of such standards should be given highest diligence and cautiousness.

The fundamental question is: Who has developed the standards, based on what fundamentals and experience, and who is responsible? Participation in working groups on norms and standards is often an honorary, unsalaried job, or reimbursed on a very poor basis. Rules and standards are usually a compromise to suit all participating stakeholders, and often the interests of those with a strong financial background prevail. Only few group members can afford not to be paid by companies with an economical interest. Such interests are usually not about benefits to the public, as they pretend to be. Technical standards rarely represent best solutions, are usually not continuously developed, and even then would always be a step behind the actual state of the art. Often, such documents have been developed without taking into account practical experience. The authors do not bear any responsibility for the implementation. Examples of consequences from inapt standards in road tunnel safety are described in chapter 11, but inapplicable rules worked out by incompetent theorists are common in many other areas of expertise.

In my professional environment, many technical questions can be answered by applying simple analysis, based on a few important arguments. Unfortunately, such an approach is mostly not accepted by the responsible political bodies. Requirements from design rules and guidelines are to be strictly applied, often resulting in absurd conceptual and technical solutions. An expensive, time consuming Quantitative Risk Analysis is worked out, which often comes to the same conclusion as a simple logical deduction. Relying on rules and standard operational procedures, instead of thinking and using their own mind, leads professionals to take risks that they don't really understand. This usually works well for standard situations that are covered by the rules, but when facing unexpected problems, those professionals are quite helpless, because they did not learn to think and solve problems on their own.

According to my own observations in many countries with different levels of safety legislation, restrictive rules and regulations encourage people to rely too much on standards to which they transfer their personal responsibility and neglect risk awareness, reflecting and reasoning. Engineering based exclusively on standards prevents finding good solutions. Rules serve as anchors, particularly for people who are not familiar with the matter from previous experience. By bad technical standards, young professionals are conditioned to dullness at the beginning of their career. In daily engineering work, the knowledge of rules and standards seems to have become more important than real technical and systemic understanding. The effect of taking away responsibility from experienced practitioners and putting in charge bureaucrats who know little more than poorly conceived standards and guidelines, is devastating.

Analytical thinking and the solving of technical problems is suppressed in order to follow rules and standards. Engineering is replaced by bloated bureaucracy. Working as a designer, you often have to deliberately apply technical solutions that are inefficient or obsolete, and often don't even work, but are in compliance with contractual terms and standards that are mostly out of date. Do such circumstances provide incentives for enthusiastic engineers to do responsible work?

A better approach would be if rules and standards did not prescribe technical solutions, but define goals, performance criteria and how the achievement of those goals is to be verified. The implementation should be left to the creativity and experience of the practitioners. Requirements that cannot be validated or tested out are quite useless.

## 7.3 Responsibility and Liability

*'If a builder build a house for some one, and does not construct it properly, and the house which he built fall in and kill its owner, then that builder shall be put to death'*

(Code Hammurabi, one of the oldest known legal texts, approximately 1780 BCE)

In the times of the old Babylonians, liability was strict, and engineers and builders were clearly responsible for their work. The law provided a strong incentive to minimize risks in the own interest of the builder. However, building codes are not everything. As we know today, the Babylonian empire did not last forever, so obviously some risk finally caused its decline.

Even the most prudent professional is subject to errors, unexpected side effects and random events. Strict enforcement might lead to nobody wanting to take responsibility. Therefore, limited liability is one of the basic fundamentals of any economic system and technical progress. Limited liability means that part of the risk is taken by a community in the form of insurance. Any technical plant must be covered by a liability insurance, both during construction and in operation. The risks that are not accepted by a private insurance company are finally left to the state, i.e. the society. In turn, society imposes safety rules, should control their fulfillment, and often finances the necessary measures. But the role of the legal system is ambivalent. Liability acts and a strong legislation enforcing safety standards have increased safety without doubt.

However, there is an important aspect why legal frameworks also may increase risks in practice. Most technical solutions rarely represent the optimum regarding usefulness, benefits, costs and risks. Technical equipment often doesn't work or is obsolete. However, from a legal point of view, the investors or operators who follow a conservative approach, relying on the existing concepts and solutions, rarely do anything wrong, as long as the flaws are not too obvious. This prevents improvement and progress. Decision makers who must consider the possibility of being accused in case of an incident related to their decision, tend to be extremely conservative and stick to the standards and rules, no matter how useful such standards are. But then, what happens when a standard is obviously obsolete from a technical point of view, even when still legally binding?

Responsible professionals should strive for a high level of quality and safety without being obliged by outdated rules. A new technical solution may be much safer than a concept that is in accordance with the relevant standards. Such standards often require useless or even counterproductive measures. However, if an incident happens, followed by a legal investigation, the judges and lawyers consult whether the technical rules and standards have been followed. Legal issues are not about good technical solutions, but about abstract paperwork, which is often contradictory to simple logic.

Consider the following example: A technical system is designed according to a thoroughly elaborated concept, but not according to actual standards, and an incident leads to one fatality. Then the designer will be sentenced for not following the rules. If the same incident happens in a similar technical system that is in accordance with the standards, killing 10 people, then the responsible engineer is acquitted. In any legal proceeding after an incident, the people who have worked out the standards don't bear any responsibility. Lawyers don't understand technical issues anyway. By transferring responsibility to rules and standards, responsibility is in fact suspended. It is quite the opposite to the strict liability laws of the Babylonians. For instance, if an analysis about a new technology shows a benefit and good business for the next few years, but possible high risk with manifestation of damages in the long run, what decisions are to be expected from executives that are hardly ever blamed for their past decisions, and not at all after they are not in charge anymore?

## 7.4    The Missing Door

Emergency exits serve as protected areas in case of an incident, particularly a fire in the tunnel. The exits must be separated from the tunnel by emergency doors. Their purpose is to separate the exit from the traffic space so that in case of a fire in the tunnel, flames and smoke cannot protrude. Such doors have to meet many requirements, for instance compliance with fire resistance norms, and all this must be filed as part of the safety documentation. Most importantly, people should be able to open the door in case of an emergency. This may be a problem under dynamic pressure conditions in a tunnel, taking into account meteorological boundary conditions, pressure surges by passing cars and trucks, and possible operation of the ventilation equipment. In fact, tests and real emergencies often show that the emergency doors could not be opened, unless by a strong athlete. Many road authorities solve that problem by simply omitting the tests.

Once we had to assess the safety documentation of a road tunnel in a country that is notorious for its excessive bureaucracy. Many office workers produce tons of paper, focusing on compliance with guidelines and rules and standard operating procedures. While checking such safety documents, we had the brilliant idea to assess the equipment not only in the documents, but also in the tunnel. When we wanted to have a look at the emergency door, the door was missing. It had been temporarily removed for refurbishment, but the tunnel operators did not even know about that. In case of a fire, the smoke would be blown from the tunnel into the emergency exit, setting a deadly trap to the escaping people. However, according to the safety documentation the door fulfilled all requirements.

Fig. 21    Emergency exit without door

# 8 Psychology

## 8.1 Mountain Adventure

My experience with the slab avalanche, as mentioned in the preface, is an example of taking a limited risk due to decision making under uncertainty. The day when the incident happened I was aware of the avalanche danger. When I set out to explore the landscape with fresh snow, being very cautious in finding a way for a safe ascent, I finally chose to cross a short slope that might be possibly dangerous, but became flatter downwards. Therefore, I considered the risk as acceptable. My consideration proved to be partially right.

When the slab broke loose, it was not completely unexpected. The movement was slow, nevertheless it dragged me down, and the feeling of complete help-lessness overwhelmed me. My lower part was buried in snow, but luckily it was not a threat to my life. As I had expected, the avalanche stopped after a few meters. Within minutes, I freed myself from the snow and returned to the place where my friends had waited. My sunglasses and a ski pole stayed buried in the snow pile, and were recovered the next summer. In all honesty, that little incident had been rather frightening than really threatening, but in hindsight, I had neglected the residual risk.

Fig. 22    Slab avalanche in which I had got caught

Did I really assess the limited risk realistically, or was I simply lucky? As a thought experiment, what would have happened if I did exactly the same thing a hundred times, but with slightly changing boundary conditions? For instance, if the snow layer would have been only a few centimeters thicker? If there had been a hole in the snow, with my foot trapped by a branch of a bush beneath the surface[24]? If the slab would have been released a few meters higher? In all those alternative histories, how many times would I survive, taking into account that I was alone, without avalanche beacon, my friends being far away?

Even professional mountaineers may get caught up in avalanches, sometimes with a fatal outcome. The factors that lead to wrong perception of avalanche dangers, as described for instance by McCammon [83], are basically the same as for other risks, which are for instance overestimation of one's abilities to assess the situation, and social group behavior. This will be explained more in detail in the following chapters.

## 8.2 Risk Perception

Risk perception is not directly related to objective risks as calculated in probabilities and consequences, but prone to many subjective factors. Human risk perception is about 'feeling safe', not about 'being safe' according to objective, measurable criteria. Some feelings obviously mislead us, particularly concerning technical issues where appropriate risk awareness could not have developed in a long evolutionary process.

- Risks that are taken voluntarily are accepted more than risks that are inflicted by other people or external forces.
- Risks that are perceived as controllable are preferred to those that you cannot influence. You feel – erroneously – safer while driving in your car than as a passenger in a commercial airplane. As is explained in the next chapter, people overestimate their own abilities.
- Risks resulting from new technologies are feared more than risks from existing and proven technologies. People fear nuclear radiation, but accept air pollution and carbon dioxide emissions by coal-fired power plants.
- Risks where the damage occurs immediately are perceived as higher than those with a delay between cause and effect. Smoking is a good example of a risky behavior where the negative consequences may take effect years or even decades later.

_____

[24] This has really happened to me on another occasion

- Risks with damages that can be repaired are preferred to those with irreversible damage.
- Risks that are actually covered in the media, or discussed in your social group, bother you more than those that you rarely hear about.
- Despite of risk being by definition the product of probability and consequence, the focus is set on consequences, but probabilities are strongly neglected.

## 8.3    Feeling and Reason

The concept of humans behaving as rational agents has been abandoned among economists after extensive scientific research in behavioral economics, cognitive and social psychology. Engineers still have to learn that. Many comprehensive and highly recommendable books were written on those issues, for instance by Kahnemann, one of the originators of the prospect theory about decision making under uncertainty, and other brilliant scientists. Kahnemann describes the human brain as operating with two different mechanisms, in particular when analyzing a situation, solving a problem or making a decision [67]. One is automatic and fast, relying on feelings and heuristics, the other is strenuous and slow, based on conscious reflection and systematic approach, which is what we generally define as reason. In our brain, there is a permanent cooperation, and sometimes struggle, between feelings and reason. However this is only one of the rough simplifications that enable us to understand complex issues. Heuristics and biases serve as a shortcut to profound analysis in decision making. McCammon writes in regard to avalanche accidents [83]:

*'The heuristics are fast, convenient and most of the time don't result in accidents. In contrast, knowledge based decision tools are often slow, tedious and can yield ambiguous results.'*

Fast processing based on feelings has developed in animal evolution as a form of 'natural' risk management. Logical reasoning is a new feature, probably used exclusively by humans and a few other higher developed animals. Obviously, in nature and traditional societies, complex analysis and the cognition of probabilities were mostly not necessary. The average human brain is not configured for extensive rational and logical thinking. In fact, logical reasoning is an exceptional state, which requires deliberate effort. Humans think and act mainly guided by their feelings and emotions. As Damasio has described, without emotions we would even not be able to make any useful decision at all [27]. In particular, women are well known for relying more on feelings, but in fact, men seem not to be much different. Brilliant logical analysts are the exception among both genders.

But generally speaking, most women tend to emphasize feelings and social issues rather than logical reasoning, and in return perceive application of strict rationality as a form of autism. This seems to be one of the basic roots of mis-understanding between the genders. However, regardless of gender, emotions are essential to explain risk related human behavior. But emotions change, and so does risk perception. In the following pages, a brief overview of the most important biases and heuristics is provided. For details and examples refer to the bibliography in chapter 14.1.

**Availability Bias and Representativeness Heuristic**

Availability bias leads to the overestimation of probabilities of events that come easily to our mind. Humans tend to assess risks based on small random sam-ples of easily available ideas. This explains the repeating cycle of high risk awareness after an incident followed by increasing carelessness until the next incident happens.

We think that 'what we see is all there is' and don't take into account that the vast majority of facts and possibilities stays hidden. That does apply to all is-sues of work, science and life. Humans always overestimate their knowledge. The quality of the data on which conclusions are based plays a key role. Basis data for risk analysis are not always derived from a serious scientific approach, but rather on rough assumptions and suggestions.

By using representativeness heuristic, people judge different risks and their possible causes and probabilities by their similarity to ideas and prototypes, which they have got actually in their mind - the 'typical' case. People form their individual opinions based on their colleagues' opinions, second-hand infor-mation and actual media content, without questioning the quality of the data and trustworthiness of the sources. Availability and representativeness are usually not correlated to usefulness, but determined by personal experience and random information from our social environment. They are stronger than objective information about probabilities. Small risks are either completely ig-nored or exaggerated. Notably the media, providing sensational stories and pictures, seem to cause exaggeration by an availability cascade in a self-reinforcing cycle [76]. In fact risk perception in the media is quite inverse to reality.

Availability bias and representativeness heuristic can be useful in the form of expert intuition. It is important to distinguish the relevant from the less im-portant. To experienced professionals, when a decision has to be made facing a potential risk, similar situations from their fund of experience become availa-ble. However, when the risk is new or unique, there cannot be any experience.

## Anchoring and Adjustment

In any decision making process, the first value that is presented to our brain serves as an anchor for any further considerations, for instance target numbers in a planning process or probabilities in any risk consideration. Surprisingly, this works even when we know that the anchor value is completely wrong or has nothing to do with the issue to which it is related. Kahnemann [67] and others describe many experiments from social psychology that show this effect with spectacular results, for instance when people have to give an estimation.

Rules and guidelines are especially important in that respect, because they serve as anchors in any professional endeavor, and are rarely ever questioned. Another example for anchors are extents of damage, where the worst case that has happened so far is considered as an absolute 'worst case' and perceived as a benchmark for risk perception. Not only are worse incidents beyond our thought, but the incidents with less damage that might be much more probable and therefore should bother us are strongly neglected.

## Affect

Most of our decisions are worked out unconsciously, and our brain invents a logically reasonable explanation only afterwards. Feelings are related to availability and representativeness, leading to occasional overreaction to specific and immediate risks – whether they are real or only imaginary. But feelings are also responsible for the negligence of abstract risks with delayed consequences.

Humans are conditioned to 'freeze, fight or flight' when encountering a threatening situation[25]. When facing immediate danger, logical thinking and problem solving are critically impaired. In such situations, people get upset, panicked, frustrated or angry, and human behavior may become extremely irrational. This may cause fatal errors, for instance in the case of the AF447 passenger aircraft crash. In subsequent trials after incidents caused by particular people, affect is often not understood, and culprits are judged against expected rational behavior, which is simply non applicable.

---

[25] Often reduced simply to 'fight' or 'flight', which is an improper simplification. Freezing is the most often observed reaction, particularly when facing an unknown threat.

## Simplification

Systemic risks in ecology, economy and society are complex beyond human comprehension. Even technologies, which have been designed and realized by humans, achieve a level of complexity that is not readily understandable. Simplification is a human strategy to understand a complex world[26]. Without simplification we would not be able to think or act.

Unfortunately, by simplification we blind out many aspects, which would possibly be important. When the question is too complex, we tend to answer a different, but simpler question, often replacing logical analysis by feeling.

A typical example are engineers facing the task of evaluating a technical solution for a given problem. Instead of answering the question: 'How to solve the problem?' they often answer the question: 'What is written in the guideline?' The difference is subtle, but crucial.

Risk is in the details, and simplification can be ominous. By simplification, many important aspects might be disregarded. Systemic risks arise when we base our decision on wrong models, which would be appropriate for simpler issues. Reality is always more complex than we imagine. The reduction of complex risks with stochastic characteristics to simple deterministic relations of cause and effect may lead to fatal underestimation.

## Premature conclusions

People derive conclusions based on little randomly available information. Probabilities are derived from a (too) small number of examples. Erroneous evidence is based on poor statistical data, without questioning its significance, sometimes only on a single random event. Particularly in safety engineering, this is a very common bias, because many incidents usually don't happen in statistically significant numbers, and there is a tendency to regard those few examples as authoritative cases.

A common premature conclusion is the deduction of a cause and effect relationship from random statistical correlation. Where there is simple chance, we misinterpret a meaning. Forming logical histories from rough data is one of the key working mechanisms of our brain. When the history sounds good, the quality and quantity of the underlying data are not questioned. The less you know, the easier it is to construct such a history.

---

[26] In contrast, some professionals like for instance lawyers and consultants profit from making simple problems complex.

## Hindsight Bias

People overestimate the predictability of past events once they know how they turned out. One is always wiser with the benefit of hindsight, and retrospectively anything looks predictable. We think we understand the past, i.e. the reasons that led to certain events, and from that we derive rules to predict the future. From data that are acquired after an incident, a logical story is constructed. However, such data were not available previous to the incident. Probably the real story was different, but it will stay unknown. Rearward risk reduction, when it's too late, is common practice. After a catastrophe, many wannabe experts know exactly the reasons that caused the incident and how it could have been avoided. Any failure can teach us a lesson on how it could be avoided in the future. But often, the causes have been misinterpreted and not really understood. Many disastrous incidents are unique, and there will be no second time to apply the lessons learned.

Past events are mostly perceived as deterministic, because when we know the outcome, we construct a cause-effect relation even for pure stochastic processes. We always need a logically sounding explanation, and preferably someone to blame, especially in legal proceedings.

## Disregard of Dead Witnesses

An important aspect of Hindsight Bias is the disregard of alternative histories and 'dead witnesses', leading to underestimation of risks of events that have been luckily survived. For many risky endeavors, a few survivors bear witness, but the many fatalities disappear and stay silent. People who succeed taking high personal risks are considered as successful, but the majority that has failed, having taken those same risks, is not there anymore. If something has worked for you, that is no guarantee that the risk was low. Maybe you were just lucky. If it hadn't worked, you wouldn't be there to question that.

I consider myself a good example for this issue. After living a childhood as a coward, I wanted to prove my courage in my youth by taking many unnecessary risks in adventure sports and violent circumstances, providing sufficient material for another book. In case of doubt, I did it. My friends perceived me as a lucky guy, and occasionally I felt invincible. But then, a few times I fell flat on my face or got my butt kicked. The insight that my success was not based on my abilities, but rather subject to pure luck, was enlightening. Sadly, some of my friends will never write about their adventures, because they did not survive their wild times.

## Overestimation of Abilities

People tend to hold overly favorable views of their abilities in many social and intellectual domains [74]. In any survey about self-assessment, approximately 90% of people consider themselves as having higher-than-average capabilities in many areas (professional abilities, risk assessment, financial performance, making forecasts, car driving, any personal skills ...). If you understand what 'average' means, that would be mathematically possible only if the remaining 10% performed abysmally.

Classic examples of overestimation are ambitious project schedules. In reality most projects take longer and cost more than expected, according to the so-called 'planning fallacy'. Self-confidence, overestimation of capabilities and planning fallacy are rewarded in a competitive environment where you have to be better and cheaper than your competitors, but it lets you take higher risks. The avoidance of a disaster is often the result of pure chance, but afterwards we see it as our contribution. That increases our self confidence, but narrows risk awareness.

## Confirmation Bias and Avoidance of Cognitive Dissonance

In decision making, various arguments, which often are not compatible to each other, must be taken into account. This is contradictory to our desire for logical simplicity: 'What must not be, cannot be'. Therefore we prefer only information that is in accordance with our preconceptions. Rather than facts driving beliefs, our beliefs dictate the facts we chose to accept. And those beliefs are driven mainly by feelings. For any issue, there are many contradictory arguments, and every alternative has its own inherent benefits, risks and costs. But this is perplexing. Our brain is adapted to a simple world of good and bad. If we have to decide between two variants, both with benefits and disadvantages, we get into a mental struggle.

For variants, which have their advantages in a certain preferred context, humans tend to ignore the disadvantages, and vice versa. Good technologies must be without risks, and risky technologies are bad. People always look for arguments that confirm their opinion, and neglect the contradictory, setting higher standards of evidence for arguments that go against their current expectation. From the information overload that is available today, mostly on the Internet, it has never been so easy to confirm any preconceived opinion, regardless of correctness and trustworthiness of the underlying information.

Confirmation bias leads to a continuously stronger belief. Many disasters happened because people felt being 100% right, even when all objective evidence showed the contrary. They had focused on a small part of information that fit into their preconceived opinion.

Often we are blind to the obvious, and we are blind to our blindness, since the problem about blind spots is that we don't even know they are there. Deliberate blindness is applied in the form of the 'ostrich effect', the avoidance of apparent risks by pretending they do not exist.

*'We see what we expect to see and are blind to the unexpected'*

## Backfire Effect

The backfire effect or belief perseverance might be seen as a natural defense mechanism to avoid cognitive dissonance. Rational arguing and presentation of facts do not necessarily convince people. In contrast, daily life and social experiments [95], [75] show that many people stick to their wrong beliefs even stronger after their arguments have been proven wrong by facts.

Giving up a preconceived belief is perceived as a kind of loss. We identify ourselves with our beliefs, and are emotionally tied to our convictions. Any arguments against them are perceived as a personal attack. Therefore, any doubt is suppressed. Even when recognizing that we have made an error, we have problems to change our behavior. Instead, we do things the same way as we have always done before, although knowing that it's the wrong way. Such behavior seems to be especially common among decision makers in charge, who simply are not able to admit errors, being afraid of losing face. Unfortunately, not recognizing mistakes in risk issues may have disastrous consequences.

## Loss Aversion and Uncertainty Avoidance

Losses loom much larger than corresponding gains. Failures are difficult to accept. Uncertainties are feared as risks, rather than opportunities. This leads sometimes to irrational conservative decisions. Many people prefer something bad, but familiar, to something new, which would be most probably better. That may be a reason why for instance road tunnels had been equipped for many years with conceptually inapt ventilation systems, which in case of fire even increased smoke propagation. Public servants tend to be extremely reluctant to try out something new, even when the old obviously doesn't work. They cannot be personally blamed for the flaws of the technologies that they have taken over from their predecessors, but any improvement bears a risk of failure that they would not take.

In contrast, under circumstances of certain loss, people tend to increase their readiness to assume risk. If you have nothing to lose, safety is not so important anymore, which leads to the proverbial courage born out of desperation. In wars, this has been applied successfully by many commanders who literally burnt the ships, depriving their troops of any possibility of withdrawal. A famous example is given by Hernán Cortés, who landed in 1519 on the Mexican coast and faced a mutiny by some of his men who had become anxious confronting a mighty alien power. After burning their ships, the Spanish had no other chance than to march forward, and finally they conquered the Aztec Empire.

## Halo Effect

As we know from politicians and sales people, handsome and confident speakers with good rhetorical skills will more probably convince their dialogue partner, even if their way of arguing is completely wrong from a reasonable point of view. Famous actors and sportsmen are regularly interviewed in the media about their opinion about actual political and social issues, and perceived as opinion makers, although it's quite obvious that their competence is in other areas.

In the engineering business, that's not different. Experts and consultants are chosen for rhetorical abilities or sympathy, rather than for proven professional expertise, and often they don't really understand what they talk about. Neither do their employers and clients, mostly executives and managers without any technical background. This explains many absurd decisions in risk issues.

## Group Leveling, Peer Pressure and Conformity

Decision making in groups is a fascinating field of societal and psychological research. Humans are social animals. In traditional societies, belonging to a group was essential, and social exclusion would impose a deadly threat. Therefore, peer pressure and human strive for affiliation is inherent in our nature. Unfortunately, peer pressure and group consensus explain many failures and disasters due to wrong decisions, even causing deliberate harm.

In a group, risk perception and decision making works differently compared to single individuals. Being tied to a group makes you feel good, and for that feeling people are prone to give up reason. Due to unconscious psychological processes, group dynamics and the bystander effect, the urge to comply with group standards becomes more important than logical reasoning. Even more under conditions of uncertainty, people prefer to follow the opinion of others.

People choose groups in which the other members have similar thoughts and attitudes. By that, they narrow their mind, and sometimes develop strange, irrational opinions, especially when isolated from other social groups. A whole group may neglect significant risks that would be obvious for outsiders. Well known examples about fiascoes caused by erroneous groupthink have been examined for instance by Janis [64] and some examples from my professional environment are described in this book.

Group dynamics is about conformity, social consensus and 'feeling good', not about reason. Discussions in groups, even among competent professionals, are often a matter of conviction instead of rationality. Group leaders are often fast thinkers with rhetorical abilities and social skills. The issue is about power, manipulation and enforcing one's opinion. It doesn't matter whether that opinion is objectively right or not. Rhetorical abilities become more important than proficiency. Under such circumstances, decision making is rather a question of prevailing prejudices and personal interests than of finding solutions by thorough analysis. Only few decision makers have got the proficiency to admit errors.

In contrast, most group members don't even make any attempt for independent analysis, but simply follow the others, since this is more convenient. Pluralistic ignorance is the rule, which states that a majority of group members privately reject a norm, but incorrectly assume that most others accept it, and therefore go along with it. Pressure of maintaining group consensus directly results in the lack of independent thinking and analysis. For group affiliation, people speak and act against their own conviction, and adapt their opinion to group consensus, even against objective evidence, as was demonstrated by Asch [3] and Sherif [121]. An important aspect is that belonging to a group is achieved by being different from others. That is not only revealed in the fashion of teenagers, but also in beliefs of adults, particularly in the engineering world. How else can it be explained that contradictory technical requirements are worked out by different experts to achieve identical goals, considering that the same physical laws apply?

Kahnemann proposes the following method to overcome peer pressure in discussions, at least in the first stage: Write down your contribution and opinion, and hand the sheet to all other participants, before the discussion is opened. I tried to apply that successfully in my professional environment even before having read Kahnemann.

For other aspects of group dynamics and social issues in organizations, see chapter 10.

## 8.4    Risk Compensation

Risk compensation (or risk homeostasis) was first described by Peltzmann, who investigated driver response to automobile safety regulations [96]. Safety measures make people feel safer, but feeling safe may encourage them to take higher risks. As a very simple model, people are used to a certain, strongly subjective risk level, where they feel comfortable in their daily life. When a threat exceeds this level, people will modify their behavior and become more cautious to reduce their risk until they are comfortable again. But when the subjectively perceived risk level drops far below their usual level by safety measures, they will again modify their behavior by increasing their level of risk until they are once again in their target zone. Thus, the psychological effect of 'feeling safe by being protected' has the drawback that it might lead to carelessness. Many examples from road traffic have been investigated, where risk compensation has a significant effect, for instance seat belts, airbags and brake assistance systems (ABS) in cars, and bicycle helmets [138]. Having an airbag in your car doesn't mean that you shall drive faster. An important aspect is that aggressive driving habits, encouraged by vehicle safety measures, increase the hazard to other traffic participants like pedestrians and cyclists.

However, risk compensation is controversially disputed. To which degree risk compensation really may reduce the effect of safety measures is subject to particular conditions and individual behavior.

## 8.5    Open Windows

Closed rooms need to be ventilated to ensure an adequate air quality for the presence of people and sensitive equipment. You don't need to be a ventilation engineer to understand that a fast, thorough airing is provided when two windows on opposite sides of a room are opened. There are always natural pressure differences between opposite sides of buildings, which cause airflow through the openings. Many buildings are equipped with windows that can be fixed in a slightly open position. With decreasing opening area, the ventilation performance deteriorates. By fixing the opening angle, the ventilation rate can be roughly adjusted. With only one open window, and no openings on the other sides of the room, the ventilation is quite limited. Many people without technical education understand and apply this basic principle unconsciously.

In moderate and cold climates, a lot of energy is used for the heating of build-ings in winter. Almost everybody claims that she or he is in favor of doing 'something for the environment' in times when the negative impact of energy waste, particularly from fossil fuels, becomes obvious to the general public. But then, when you visit the same people, you see that they let their bathroom window permanently slightly open in tilt position during the whole winter. The desired effect – ventilation – is hardly achieved, but the negative effect – heat loss – is considerably large. In other words, they waste the heating energy literally out of the window. There are many advisory leaflets that describe how to provide effective airing in winter by periodically applying a brief and intensive draught. For many years, I tried to explain it to other people, with little success. Most nodded, but still kept their windows open, showing a vivid example for the persistence of bad habits and their hidden negative consequences.

Fig. 23    Slightly open turn / tilt window

Since it is very difficult, if not impossible, to change bad habits of people, engi-neers invented modern low energy buildings in which the windows cannot be opened at all, and forced ventilation is provided by equipment that facilitates energy recuperation. This is a good solution from an engineering point of view, but does not take into account human psychology. To most people, including myself, there is a fundamental need to open a window and breathe fresh air directly from outside. Moreover, permanent but reduced ventilation through slightly open windows is useful in summer and in regions with warm climate. In cold climates, turn / tilt windows seem to be one of those technologies that mankind is not mature enough to apply.

# 9   Decision Making

## 9.1   The Right Decision

To our ancestors, the world was mostly determined by unswayable forces, which they did not understand. Their decisions had little influence on the risks that they faced. With todays scientific understanding and medical and technological progress, our decisions have a significant impact on our safety and wellbeing.

Decision making is a permanent, mostly subconscious process of daily life. Our decisions may be questioned by ourselves and by others: Have we made the 'right' decision'? Unlike in mathematics or other well-defined constructs of ideas, in real life there is no absolute right or wrong[27]. Everything has advantages and disadvantages, depending on the point of view and momentary framing conditions. Life is a dynamic process of evolution and development. What is considered right today may be wrong tomorrow. Any endeavor bears the possibility of failure and even disaster. In a complex world, decisions have to be made under uncertainty, between different variants with particular benefits and disadvantages. Unfortunately, many decision makers operate with a children's sight of the world, with a simplified concept of right or wrong, good or bad.  In real life, contradictory requirements have to be met, taking into account different advantages and disadvantages, which are often a question of point of view. There is very rarely a simple decision between risk and safety, but usually between an option with particular benefits, costs and risks, and another option with other benefits, costs and risks. Which one we prefer is rarely a matter of objective weighing of factors.

Pros and cons are often difficult to assess and a thorough decision making process might be infinite. This is well described in the tale about the donkey that starves to death between two equal heaps of hay, not being able to decide on which side it should start to eat. An example from reality is the use of DDT, an insecticide used for disease and pest control in and after World War II, and its ban in most western countries in the 1970s after the revelation of massive ecological issues caused by its application. For instance, DDT caused eggshell thinning of birds of prey like eagles and falcons. Breeding became impossible, which almost led to their extinction.

---

[27] In this book, the term 'wrong' is applied referring to undesired outcomes with negative consequences.

However, such ecological problems make us blind for the very benefits of DDT in fighting infectious diseases and crop failures caused by insects. Costs, benefits and risks are very difficult to weigh against each other. Typhus and malaria could probably be extinct by excessive application of DDT, but what would have been the consequences of its ecological impact in the long run?

Knowledge does advance, and circumstances may change. The consequences of decisions are never fully foreseeable, since positive and negative surprises can occur that nobody has thought about before. It's easy to be smart with the benefits of hindsight. Even when we think that our decisions are based on logical reasoning, a large part of arguing takes place on a subconscious level. 'Intuitive', fast decision-making methods based mainly on feelings were perfectly adapted to the natural, traditional world. However, for specific technological risks, quantitative methods, in particular Quantitative Risk Analysis, would be more appropriate.

For rational decision-making, first of all clear goals should be defined. Naturally, there are many possibilities to strive towards such goals. Those possibilities lead to different variants of concepts and technical solutions. Each variant has its advantages and drawbacks, an optimal range of application, limits of adaptability, uncertainties and costs.

For complex issues with fundamental significance, solutions should be evaluated by a systematic approach, examining all useful variants. That may be a useful application of Quantitative Risk Analysis as part of a profound cost-benefit analysis. At the end, what counts is whether a concept or a system achieves the defined goals or not, under which conditions and at what price, taking the whole lifetime into account. Whatever your decision may be, you must accept the resulting drawbacks, costs and risks.

A proposal for a good decision making process would be as follows:

- Define clear goals that are easily verifiable.
- Work out different concepts to achieve those goals.
- Gather necessary information as far as possible, assess its trustworthiness and take into account that there is an unknown part.
- Thoroughly analyze each concept in terms of benefits, costs and risks.
- Take into account both realistic and extreme scenarios, estimate the consequences, assign appropriate probabilities, and be aware that there are unexpected hazards that nobody thinks about.
- Examine the influence of parameters by a sensitivity analysis

- Discuss the issue with competent professionals in an atmosphere of mutual respect and open communication, considering both scientific knowledge and practical experience.
- Base your decision on an evaluation according to predefined, clear criteria, but listen also to your intuition.
- Implement the concept and measures according to the decision. During that process and after completion, review whether the goals are still useful and whether they have been achieved or not
- Test thoroughly whether it really works, on a regular basis during the whole lifetime. Testing must be feasible and not lead to unacceptable damage.
- If goals are not fulfilled due to changing conditions or deterioration, apply appropriate measures for change to ensure that actual goals are met.

Unfortunately, reality is different. Nobody would invest the time and resources for such an idealized process. What I see in my professional environment as well as in society in general are the following issues:

- Goals are not clear, often not verifiable, and sometimes not even defined.
- Information is always incomplete and distorted.
- Only one or a few, randomly selected, at first sight most convenient concepts and technical solutions are taken into account.
- Benefits and risks are not really understood. 'Dread' is overestimated, probabilities rarely taken into account, and abstract risks are neglected.
- Lifetime costs are underestimated or rather completely unknown.
- Discussions are mere rhetorical banter where preconceived opinions are exchanged. Opinion leaders with a big mouth, but little professional competence, assert their interests.
- Commercial arguments prevail, disregarded of the usefulness and risks of proposed solutions.
- Conceptual, educational and organizational solutions are neglected in favor of technical solutions with business opportunities for particular stakeholders.
- After completion and payment of bills, the whole issue is forgotten. Maintenance and regular testing are ignored. Failures are discovered only when a serious incident happens.
- A few profit when everything is going well, but nobody is responsible when things go wrong. In that case the society has to bear the costs.

## 9.2    Mindfulness, Awareness and Alertness

Mindfulness and awareness are important factors in risk management, whether as situational alertness in case of possible immediate threats, or in the form of careful analysis and decision making in a planning process. Being able to think with a clear head is essential, but exhausting, since mental capacity is limited. Logical thinking and decision-making is hard strenuous work, requiring mental energy. So is a state of alertness and deep concentration. The term 'pay attention' must be understood literally: Attention is not unlimited and must be paid in terms of time, energy supply and recuperation.

Factors that reduce awareness and the capacity of logical thinking, affecting the ability to be vigilant and to concentrate on risk avoidance are mainly the following:

- Fatigue and exhaustion
- Sleep deprivation
- Intoxication (by alcohol or other drugs)
- Weakness, sickness and illness
- Social pressure and expectations
- Time pressure
- Distraction and multitasking

Mental exhaustion reduces our ability for self-control, which is an important performance factor. Depletion of mental resources significantly increases human error. A worker on an undemanding routine job may spend more than 12 successive hours working, even when with decreasing concentration the reject rate in a production process would increase. However when consequences might lead to disasters, this should not be acceptable. An executive in a responsible position, who spends 12 hours in his office, half of that time in a state of mental exhaustion, may cause great damage when making decisions. That's what happens in reality. Many professionals – not only those involved in important decision processes – are in a permanent state of distraction, kept busy by administration and paperwork, meetings and social issues. Some strain their minds additionally by intellectual junk food in their spare time. Under such circumstances, awareness, reason and mental sanity can hardly be maintained. This aspect is important in explaining the fact that most intellectual workers don't act in accordance with their professional knowledge in their private life. After a hard working day, solving challenging problems for their clients, their brain is simply too exhausted for clear thinking.

In a culture in which sleep and rest is given little value, sleep deprivation is one of the most important and underestimated factors that cause human errors. For instance in the U.S. the National Highway Traffic Safety Administration estimates that 100,000 police-reported crashes are the direct result of driver fatigue each year. This results in an estimated 1,550 deaths, 71,000 injuries, and $12.5 billion in monetary losses. However, it is difficult to exactly attribute crashes to sleep deprivation, since there is no test to determine sleepiness as there is for instance for intoxication. 24 hours without sleep have the same effect on analytic capabilities and reaction time like a blood alcohol level of 0.1% (1 gram alcohol per liter of blood). The effect of intoxication on our mental capacity doesn't need to be pointed out, since drunken driving will be punished in all civilized countries. The truck driver who caused the tunnel fire disaster in the Tauern tunnel in 1999, where 12 people died, suffered from serious sleep deprivation. The one that caused the Gotthard road tunnel fire in 2001 was drunk. He did not survive his mistake, together with 10 other fatalities.

Focusing on a task is essential for efficient performance, but deep concentration on a particular task makes us blind to the surroundings and other, possibly vital information, and reduces significantly our alertness. Such a state is necessary to accomplish hard brainwork, but is acceptable only in a protected environment. Outdoors, where you might be threatened by adverse weather, hostile fellow humans, or dangerous technologies, your attention must be everywhere.

Attention is the opposite of distraction, which is another potential source of mishaps, accidents and disasters. For instance, in the U.S. approximately 6% of all car crashes happen due to cell-phone use while driving, leading to approximately 2,600 fatalities per year. In most European countries, such behavior is subject to penalty. Modern information technology and digital means of communication contribute significantly to distraction and reduced awareness, and may lead to 'digital dementia' in particular for children [124]. While working on a computer, a part of our mental and attentional capacity is inevitably absorbed by the handling of the machine. Those new technologies also foster multitasking, which is useful for computer operating systems, but not for humans. Multitasking does not really work because the brain cannot stay fully focused when jumping from one task to another. Multitasking causes people take longer to complete tasks and induces additional errors. If the attentional load increases, mental capacity diminishes [87].

## 9.3    Knowledge, Know-How and Common Sense

*'Common sense it not so common'*
(Voltaire, French Philosopher, 1694 – 1778)

'Common sense' cannot be common, since it is in fact depending on the personal point of view, cultural and educational background, and especially the social group consensus. Humans do not always do the right things, and they don't do them in the right way, even when they know better. The distinction between knowledge and knowhow is essential. Knowledge may be obtained by reading some literature and by theoretical analysis. Know-how is about how to apply the knowledge in practice and must be earned by practical, hands-on experience. Lecture provides for ideas, but only practical experience lets us really understand them. Experience means that you have made errors and learnt from them. Some children learn not to touch the hotplate by following the parents' instructions, but most don't. They have to burn their fingers to learn the lesson. The same is true for adults in a hazardous environment. Therefore, know-how is rarely to be found among theorists who spent their whole professional career sitting in an office producing paperwork. Even having the appropriate know-how does not mean that it is applied correctly.

For any technology, what counts is the application and implementation in practice, not the theoretical concepts. Many issues cannot be identified by analysis, but are revealed only in practical application. Even when a system is in operation, proper function may be based on mere chance rather than on proper design and implementation. Deviations and occurring errors may be overlooked. Whether something really works or not can only be seen when it's stressed. Therefore, thorough testing is essential for the implementation of any concept, technology and equipment.

The role of expert intuition on decision making is often ambivalent. According to Gigerenzer [49], fast decisions based on simple heuristics can lead to equal or even superior results than thorough analysis, even in a modern, complex context. Tetlock [132] has investigated predictions from experts on political issues during two decades, and compared them to what really happened in reality. Expert predictions were not better than those worked out by laymen, and simple rules and models were superior to sophisticated, complex analysis tools.

An interesting aspect is that according to Hill [52], little children younger than approximately 6 years show astonishingly high survivability in wilderness survival situations, while older children between 7 and 12 had the lowest survival rates.

Obviously, the youngest are not spoiled by society's rules, expectations and false ambitions. They simply follow their instincts. Since their sense of orientation is not yet developed, they stay in the place where they have got lost, which is often the best thing to do. In contrast, older children feel obliged to false ambitions, but lack the experience and serenity to assess dangerous situations. Like adults, they have a mental map of their environment, and they try to bend that map by taking shortcuts. But they are not self-reliant. A little knowledge may be more dangerous than no knowledge at all.

Expertise, heuristics and a logical approach should be based on a long history of trial and error in practical applications, in a relatively steady environment, where equal actions lead to equal results with immediate feedback. There cannot be experience in something completely new. However, even when learning from experience, misinterpretation and feelings can deceive you, and wrong conclusions might be derived. Professional intuition may be prone to faults. Experience can make you too self-confident, careless and blind to the unexpected. Having overcome a dangerous situation might be an approval for your abilities, but it can also be subject to simple luck. Some of the worst disasters might be unique events, which means that learning is of no use anyway. For an individual, death is a singular experience and learning from it is impossible.

Bertrand Russell stated:

'*Even when the experts all agree, they may well be mistaken*'

## 9.4   Framing

Framing comprises a set of concepts and theoretical perspectives on how individuals, groups, and societies, organize, perceive, and communicate about the world around them. Framing is essential for risk perception. Your view of the world may significantly change risks and the efficiency of risk mitigating measures. Unfortunately, it is strongly dependent on subjective factors, like cultural and societal background, education and experience, your social environment, and even your momentary mood. Wrong framing is one of the main reasons for human error. What works in one situation, may fail under other circumstances. In most cases, the consequences of errors are negligible, but sometimes wrong framing may lead to disasters. It is essential to apply an open mindset and situational awareness, trying to see beyond the individual point of view. For the same issue, some people see only the benefits, others see rather the risks.

As citizens of affluent societies, we perceive new technologies like for instance nuclear power supply or genetically manipulated food as risky. From the point of view of most people in the world who are not so privileged, it's the benefits that count. For instance, in case of genetic engineering, benefits are improved resistance against droughts or pests and therefore lower risk of crop failure. Taking into account that to many people starvation is a real threat, any unknown risks related to the same technology are rather neglected.

Social norms are an important aspect of framing. Living in a multicultural society, with the possibility to travel to any place in the world, is a modern feature that lets us forget that other people may have a completely different view of the world. A sarcastic sense of humor may be good for a laugh with your friends, but can get you killed in other, not so humorous societies. In interpersonal violence and self-defense issues, wrong framing may lead to fatal consequences. Martial artists and sports fighters often get beaten up on the street because they have a mindset of fair man-to-man combat, which they are used to from sporting competitions. Reality is different, as I have found out while gaining a little fighting experience in the ring and 'on the streets' in my youth. Unarmed man-to-man combat does occur only very rarely in real conflicts, being rather subject to schoolyard brawls, sporting tournaments and Hollywood movies. Winning a real fight is rather achieved by surprising, deceiving and outnumbering your opponent, acting as a group in a coordinated and intelligent manner, using superior weaponry, and the application of dirty tricks. Sports fighting abilities are only of secondary importance, if at all. However, any confrontation bears an unknown risk for all involved parties.

## 9.5   Theory and Practice

*'In theory there is no difference between theory and practice. In practice there is'*   (Lawrence P. 'Yogi' Berra)

Before graduating as an engineer, I spent plenty of time working on construction sites rather than in the classroom. For instance, I once had a job as an electrician on the installation of the cooling system for the Cray-2 supercomputer at the Swiss Federal Institute of Technology, where I also studied at that time. Coincidentally a few months later, as a research assistant at the Institute of Fluid Dynamics, I developed mesh generation codes for computer-aided fluid simulations that run on that same computer. In my following professional career, I always strived both for theoretical understanding and practical experience. Unfortunately, today the engineering world seems to be strictly divided into theorists and practitioners.

Most executives show professional incompetence, according to Putt's Law, and therefore look for experts to support them. But the choice of such experts is mostly based rather on academic titles and theoretical knowledge, than on practical expertise.

As a student, I admired the previous generation of engineers that worked out calculations with slide rules, relying on analytic methods and analogue signal processing. Later, computers became the standard tool of engineering. Sophisticated computational codes were developed and increasing possibilities moved the focus on numerical methods. Today, such simulations are worked out for many applications, for instance in the form of Computational Fluid Dynamics in my field of work. The results of such simulations pretend to have a precision and trustworthiness that is not justified, since they depend strongly on assumptions of initial and boundary conditions. Rough calculations based on rules of thumb would be more useful for practical applications. Most importantly, simple models and sophisticated simulation programs should be regularly cross-checked, as well as compared with measured data from test benches and realized systems, taking into account measuring uncertainties.

The simulation guys are often perceived as experts in their field of proficiency in public opinion. However, there is a large difference between a design based on modeling and a functioning technical facility in reality. Getting something to work reliably is hard strenuous work. More important than design calculations are operational requirements, control algorithms, and particularly testing procedures to prove whether the safety goals are really achieved or not. The further away a professional is from the practical, hands-on level, the more he underestimates the effort that is necessary to realize, implement and make work a technical system and to maintain operating conditions for its whole lifetime.

Most theorists refuse to get their hands dirty. They are often brilliant scientists, but too far away from the 'real world', not taking into account the principles as described in chapter 12. They tend to get lost in complicated concepts that look great on paper, but can hardly be realized. But those expert theorists work out guidelines and conceptual requirements, often without any peer review. Inexperienced designers, relying on theoretical knowledge, but lacking practical understanding, are awarded design contracts by public tenders where the lowest bidder wins. As a reviewer, I often find that those designers work out complicated design calculations, based on highly scientific models, leading to utterly precise results with many decimal places, but make serious mistakes about relevant parameters and boundary conditions.

Many engineers do calculations rather for self-satisfaction, forgetting about what their purpose is. In system engineering, the role of the design process is often overestimated, but the operation is neglected. But the operation is what counts in practice. Design calculations provide only a boundary condition for the possible states of operation. Designers should apprehend that utter precision is useless, since acceptance tests are usually worked out with measuring uncertainties of several percent of the measured value. Design parameters have to be increased by safety margins anyway, as explained in chapter 12.5.

In an academic environment, expertise is often judged by the number of publications. Practitioners may achieve a lot of know-how, but usually do not publish. Sometimes I wonder about 'new findings' being presented by 'experts' on conferences and in papers, often about issues that have been common knowledge among practitioners for a long time.

Practitioners often get a routine in doing repeatedly the same procedure without understanding the underlying principle and physical basics. Standard problems are approached by standard solutions. With time, practitioners get a thorough experience about what works and what doesn't – as long as the boundary conditions don't change. But many practitioners also have learnt to take into account the unforeseen, by improvising and quick fix measures of problems.

# 10 Communities and Organizations

## 10.1 A Professional Meeting

Many years ago, I was participating in a project meeting at the headquarters of a road authority[28]. As responsible designer, I had to present a tunnel ventilation project, and therefore arrived in advance to prepare the presentation in the room where the meeting was to be held, according to the invitation. But when the meeting was due, nobody showed up. After a few minutes, a colleague from the design team picked me up and led me to the adjacent room, where the other participants were waiting. They had met at the cafeteria and were led to the room by the civil servant who had organized the meeting. Subsequently I was criticized for being late. After some time, another official showed up, claiming the room for another meeting, and proved the correct room number by a written invitation in his hand. We were obviously in the wrong room. The responsible public servant did neither apologize nor recognize his error, but rather insisted on holding the meeting in that particular room. The other participants only nodded. Group consensus must not be questioned, even when there was clear evidence that they were wrong.

This anecdote might be seen as a funny misunderstanding that regularly happens without serious consequences. But with some background information, things look more serious. It was a technical meeting, and all participants were highly educated professionals and specialists, but nobody dared to contradict the public servant, who finally would approve of the project as the representative of the road authority. He did not only prevail in questions regarding meeting room allocation, but also in conceptual details where important safety issues were at stake. Moreover, he had been in the board of authors of some of the new tunnel safety guidelines, even on issues that he obviously did not understand. And he consequently demanded strict obedience to those rules and guidelines, even when they led to unfavorable technical solutions, additional risks and waste of funds. The decisions of this particular public servant cost dozens of millions of taxpayer's money, hypothetical casualties, and demotivation of many enthusiastic professionals. Meanwhile he has retired, and will hardly ever bear the consequences of his decisions.

---

[28] In a 'not to be named country'…

## 10.2   Organizations and Big Projects

Humans live in societies with complex social interactions, which have been investigated by extensive psychological and sociological research. An organization is not an entity that can think, act and take risks by itself. It is humans who form the organization, make decisions and execute tasks. Therefore it is important to understand how people work in the context of their organization. From an evolutionary point of view, being part of a group might be seen as a safety measure, since your group provides you with protection and security. In traditional societies, such independent groups were smaller than modern states or companies, and a single individual could hardly survive when excluded from society. Your fellow humans protect you from external threats, but unfortunately they also may inflict violence or force you to take additional risks.

Most human endeavors require the capacity of more than just one person, leading to organizations of many different people with partially the same goal. Big projects require big organizations. Such organizations, which are able to plan and realize big projects that would be beyond the possibilities of any individual or small group, need to function reliably, preferably without friction and loss of time. Activities must be coordinated, and each member must fulfill his duty in his position for the benefit of the organization. This requires a planning authority and a system of power enforcement.

Big organizations acquire enormous power, but tend to be intransparent, inefficient and bureaucratic, and are prone to high risks. Exceeding a certain project size, the measures to achieve an adequate safety level are economically not feasible. Failures and minor incidents are common, and sometimes develop into disasters. The bigger the project, the higher the error rate is, and the higher are the efforts and costs to provide an appropriate level of reliability. Perrow has systematically investigated how complex systems inevitably lead to incidents and disasters [97]. This has been confirmed by practical experience of many professionals, including myself.

In the Bible, there is a nice story about the tower of Babel. People decided to build a tower that should reach up to the sky. This displeased to God, so he decided to confound their speech. Then the workers were not able to understand each other, communication broke down, and the construction works stopped. The same can be seen nowadays in big projects. Executive managers and practical workers obviously don't understand each other, even when they speak the same language.

Fig. 24    The Tower of Babel *(Bruegel, 1563 - Public domain)*

## 10.3   Authority and Obedience

Highly developed social animals, whether wolves, chimpanzees or humans, are organized in hierarchic structures, where subordinates have to obey their authorities. That is part of our biological heritage.

Decisions are made by some individuals and carried out by others. Established hierarchies stabilize societies and enable efficient accomplishment of tasks. Such organizations have prevailed in traditional societies facing traditional risks, and still prevail in most companies and public agencies. In societies and project organizations, decision making, accomplishment, responsibility, profit and suffering from risks is allocated to different stakeholders. The decision makers are usually not personally affected by their decisions. Those that suffer from possible risks have nothing to say.

Decision making is strenuous; therefore it is convenient to transfer this effort and the related responsibility to others. Some men like to order, others like to follow, and most want to avoid personal risks and responsibility.

Risks are inherent in any situation when decision makers (or the stakeholders that they represent) profit from ventures, but don't have to suffer the consequences of failures when the damage takes effect much later, or when responsibility is transferred and diluted. Providing such circumstances is the basic incentive of any decision maker striving for power.

Most people prefer to transfer responsibility for their actions to an authority, giving up any moral and ethic principles. Humans are prone to do deliberately what they consider to be wrong, as long as there is the excuse of simply following orders. In that case, they don't even have to ask what is wrong and what is right. As Milgram [86] has shown, approximately 75 % of people are able to deliberately impose serious suffering and harm to an unknown fellow man, when commanded to do so by a formal authority. The people participating in Milgram's experiment did so even without any further pressure and without questioning the authority of the investigators, which was demonstrated only by wearing a white lab coat.

Plato proposed 'philosophers to be kings', dividing the human race into mindless sheep and philosophically enlightened shepherds. Ideally, leaders would guide and organize their competent subordinates, carefully working out wise decisions for which they take full responsibility, fostering benefit and safety of all members of their organization. But in reality, power corrupts people.

The strive for power is one of the most important impetuses of human beings. Success in power games, social skills, and often simple chance make you climb the ladder in political organizations and big companies, rather than professional qualification. Most decision makers are absorbed by the power games to maintain their position in hierarchy and possibly climb the ladder further to the top. Such power games and intrigues prevent people from doing good jobs. Many professionals, even when starting with best intentions, get corrupted when they acquire power. Power is not about taking responsibility; power is rather about transferring responsibility to your subordinates, often without letting them make independent decisions. This increases the probability of wrong decisions and errors. History is full of examples in which inapt leaders in large, hierarchic organizations have led their subordinates to disaster. Elite dictatorships culminated in the ideologies that caused so much harm and suffering in the 20th century.

## 10.4   Peter Principle and Putt's Law

Two principles have been described, which may serve as an explanation for the many failures of leadership in practice. First is the Peter Principle [98]:

*'In a hierarchy every employee tends to rise to his level of incompetence. Every post tends to be occupied by an employee who is incompetent to carry out its duties.'*

The Peter Principle assumes that people on the upper level of hierarchy have passed lower levels where they have been successful professionals. Leaders who have achieved their position by the Peter Principle may lack leadership abilities, but they know what they are talking about, having gained practical experience in their professional career. In contrast, leaders who have not passed through the practical levels of hierarchy are often technically incompetent. This is the case in most engineering fields of activity, as described by Putt's Law [111]:

*'Technology is dominated by two types of people: those who understand what they do not manage and those who manage what they do not understand. Every technical hierarchy, in time, develops a competence inversion.'*

In a technical environment Putt's Law prevails over Peter Principle. Even when both rules are not to be taken too literally, they describe very well what can be observed in practice in most organizations:  People who have no clue are responsible for decisions with fundamental significance. Decision makers rarely have the hands-on experience to deeply understand the issues, and mostly rely on their subordinates to be provided with relevant, correct information. In case of a disaster, they are mostly not the ones who would suffer. Managers of private companies deliberately neglect risks in favor of short time profit. In contrast, government officials tend to spend excessive public funds, neglecting any sense of proportion, only for not to be blamed when something happens.

Any improvements and investments that cost time and money, not showing direct immediate benefit on a short time scale, for instance safety systems, have a hard stand. In this respect, the time perspective of high level decision makers in big organizations has to be taken into account as well. Most top executives have usually passed their 50$^{th}$ birthday, and their further career will not last much longer than another decade. Their objective is to assure their own position for the next few years and not the long time benefit and safety of any project or organization, as long as their financial assets are not in danger.

## 10.5 Whose Interest Prevails?

An important aspect of organizations and leaders is the discrepancy between individual and collective risk. There is an unavoidable contradiction between individual and collective goals, and any organization would collapse without occasional sacrifices of its members. Finding the right balance between self-interest and public wealth, individual and collective risk is subject to endless considerations and discussions.

For instance, the employees of companies have to earn money for their employers. If the primary focus of the employees would be their own wealth and safety, their company would possibly go bankrupt. Each individual has its own interests, and increasing your safety may increase the risk to your fellow human beings. Everybody wants to minimize his individual risk and maximize profit at the expense of the society as a whole. But the society are all other individuals, who behave the same way. A balance of interests has to be found. Any decision for a concept, system or equipment is a compromise taking into account different points of view from various stakeholders and thus varying weighting of arguments in a given context. Related to risks and safety, such interests may be as follows:

- The public wants its needs (e.g. energy supply, transport, communication, medical support, etc.) to be fulfilled by safe, reliable technologies.
- Company owners, shareholders and investors intend to maximize profit and minimize the necessary investment.
- Operators have an interest in low operational costs.
- In case of an incident, people shall be affected as little as possible, at least staying alive and unhurt.
- Emergency services need to fulfill their duty, that is, saving people and mitigating the damages in case of an incident, preferably without being exposed to risks themselves.
- Construction companies, equipment suppliers, designers and the financial industry are interested in complex, expensive concepts and technologies that provide business opportunities. They are not interested in proving function, performance and liability, unless forced by stringent contract requirements.
- Employees want to have a job, a regular salary and a safe workspace.
- Lawyers, experts and consultants are intrigued by hazy, contradictory or faulty requirements and specifications, which lead to endless discussions.

- Insurance companies have a financial interest in risk mitigation, even when that may obstruct other, more important goals.
- Managers, executives and politicians want to gain power, increase personal benefit, and pass on responsibility.
- Journalists need to sell stories about sensations, and therefore tend to exaggerate risk issues in the media.

Private investors try to hide risks as long as they are not personally affected. On the other hand, state officials are prone to waste taxpayers money, only for not being blamed when something happens. In this context, infrastructure projects financed by public funds are a delicate matter, since the investor is the taxpayer, who has usually almost nothing to say. The question may be raised whose interests are mostly represented, and whether funds are really used to optimize benefits and minimize risks to the public or not. Issues about deliberate neglecting of risks or wasting of funds are inevitable when taking into account that each individual follows its particular interests. A plausible explication for wasting funds on useless measures would be corruption, since there is always someone who profits on the receiving end of the funds[29].

In the case of hierarchic organizations, the lower you are positioned on the ladder, the weaker is your position, and the higher is your risk of being exposed to danger by decisions of your superiors. Everyone who has ever done military service knows what it means to be educated to switch off your brain and blindly follow orders from authorities that must not be questioned. As a recruit, you must omit independent thinking. Hierarchic military organizations serve as a model for many business companies, since they provide an efficient means to accomplish a task. But the purpose of armies is to win a war, not to foster the wellbeing of its soldiers. Your individual risk increases significantly when you are utilized as cannon fodder. Even if you want to be a hero, being happy to sacrifice your life for a 'noble cause', whether inspired by religion or ideology, there is no guarantee that your sacrifice will be worth it. Incompetent leaders may literally waste you, or your sacrifice might have an unexpected outcome. Everything you have fought for may be lost – for you it's over anyway. Unless you believe in being awarded with 72 virgins in paradise, but even that may not necessarily be pure bliss.

Traditional societies invented supernatural causes for what they could not explain. Hazards were accepted as God's will. Who would dare to question an ultimate authority that outshines all logic?

---

[29] For legal reasons I state that I don't allege that any decision maker would be corrupt.

With today's knowledge about natural sciences and statistics, we can interpret many occurrences differently, but deep in our mind, we still stick rather to our feelings. This is abused to establish power enforcement over others. In many organizations, logical reasoning is deliberately oppressed, because it might thwart group consensus, which is often the base of power for the leaders. If a statement is clearly wrong, but repeated many times without dispute, then sooner or later it may be accepted as right. Many political and religious leaders have applied this successfully. Logical thinking and reason might increase your individual risk to be exposed to ostracism and hostilities from your fellow citizens. In the Middle Ages, people who questioned the prevailing dogmas were burnt as heretics. Nowadays, in so-called liberal societies, different-minded people are silenced and expulsed from social and professional life. Sometimes it may be safer to be ignorant, at least to pretend so, to avoid becoming a martyr of free speech. You might be forced into a situation in which the only choice is either living with unnecessary risks that have to be accepted, enforced by the opinion makers and 'group think', or being excluded or even harmed by your fellow human beings.

## 10.6   Burning the Jews

In ancient times, poor hygiene was a major cause of contagious diseases. From 1347 to 1353, medieval Europe was devastated by a plague epidemic that killed approximately one third of the population. The medieval people didn't have any scientific knowledge about proliferation of germs and viruses. The diseases were perceived either as God's punishment, consequences of unfortunate planet constellations, or an evil that was caused by hostile fellow humans. In case of the latter, it became quite clear who the originator of the plague might be: There was a rumor that the Jews had poisoned the wells. In the Jewish community, the percentage of casualties seemed to be lower than among the Christian majority, even if from today's point of view, we might doubt whether the difference was really significant according to a serious statistical analysis.

There may be an explanation for an improved resilience against the plague. Jewish religious rites demanded regular hand washing and bathing. This represents a typical example of traditional risk management. Insights from the experience of many generations were incorporated in religious rules. Therefore, the hygienic conditions among the Jews were obviously better in comparison to the Christians, which were negligent with regard to personal hygiene in those times. Unfortunately, the Jews were used as scapegoats for the plague.

For instance, on the 16 January 1349, a big part of the Jewish community of Basel, dozens of people, were locked in a wooden house on an island in the river Rhine, which was then set on fire. Nobody survived. A few months later Basel was struck by a plague epidemic that killed hundreds of inhabitants. Obviously, burning the scapegoats hadn't worked. Many leading citizens had incited the hostilities following simple commercial interests, since they had substantial debts with the Jews, or saw them as rivals in the banking sector. Do you think that such a thing could not happen today?

## 10.7   Information and Communication

'*Knowledge is power*' (Francis Bacon)

Analysis and useful decision making is only possible when the relevant infor- mation is available and rightly interpreted. There are always uncertainties, and necessary information is never complete. Wrong decisions, leading to errors and possible disasters, have been based on lack of information, misinterpreta- tion of information or wrong information. Due to the problem of induction, even with plenty of high quality data, conclusions can never be made with absolute certainty. In ancient times, the acquisition of necessary information posed a big problem. In contrast, today we suffer from information overload. Obtaining and recognizing the right information for any decision process might still be difficult, if not impossible.

Most brainwork is subconscious, and so is the acquisition, processing and the assessment of information. Rough data without further analysis is worthless, and time and resources for analysis are limited. In organizations, the need to transfer the right information from the practitioners on site to the decision mak- ers in a remote office and vice versa, possibly via intermediaries, makes things more complicated. The farther away you are from the source of information, the more likely you are prone to deception. Subordinates filter the information that they provide to their leaders. This filtering process is worked out according to one's preconceptions. Information might not only be misunderstood, but also deliberately manipulated or held back. Honesty is not a common human fea- ture, even less among people who strive for power. In fact, not mere knowledge is power, but access to information and the means for filtering, ana- lyzing and distributing data. Leaders are often too remote to be aware of the practical consequences of their orders. The men with the know-how mostly haven't got the authority to make relevant decisions. Only when an incident happens, they are the ones to blame.

## 10.8  Admittance of Mistakes

A culture of open communication between all stakeholders, particularly the men on site with practical experience, the experts with extensive theoretical background and the remote decision makers who bear responsibility would significantly reduce risks. To achieve that, confidence, honesty and the ability to admit mistakes and change course would be essential for all participants. Unfortunately, such qualities are not common among humans, even less among individuals who strive for power and personal benefit. Particularly, the inability to admit mistakes is psychologically deeply anchored in us, and also a direct consequence of societal processes and legal systems. Errors are usually not perceived as unavoidable part of any process, but as consequences of incapability of responsible persons. In fact, both aspects are true, but personal responsibilities for errors may lead to stringent negative repercussions, and therefore it is mostly convenient to not admit any mistakes in a social environment. In a professional context, admitting mistakes may lead to financial claims, legal prosecution, and at least damage your reputation.

But not admitting mistakes prevents countermeasures, avoidance and progress, and therefore leads to an accumulation of risks with time. You learn more from failure than from success. Consequently not admitting mistakes is an attribute of religious and political belief systems that claim to know the ultimate truth. In Chernobyl, the inhabitants of the surrounding villages and towns were evacuated only after a significant delay, because the secretive communist Soviet government didn't want to admit that the catastrophe had occurred.

## 10.9  Financial Pressure

Any business is based on the necessity to earn money. Unfortunately, short term financial profit is not about providing useful goods and services at high quality. Measures to avoid and mitigate risks cost time and money without directly apparent benefit. Basic engineering principles to mitigate risks are described in chapter 12. But most of the mentioned measures, for instance reserves and testing, cost time and money and reduce profit. Safety systems, which are probably never needed in their whole lifetime, bind funds for investment and further expenses for regular maintenance and testing. Therefore, in the private sector there is a strong pressure to omit any safety measures. Only thanks to strong legislation in most civilized countries, safety standards have been established. But many companies save funds by outsourcing possibly hazardous business units to countries with less stringent safety standards.

Unfortunately, not all safety rules in practice really increase safety. Huge funds are wasted for measures with only diminutive contribution to risk mitigation, because there haven't been any funds for a proper risk assessment and an analysis of usefulness. Since subsequent financial pressure leads to savings in quality control, implementation and testing, those expensive measures often don't even work in practice.

Financial pressure and optimizing inevitably leads to an increase of risk. Decision makers have little understanding of underlying processes. Many investors are hardly ever aware of risks (beside the financial ones) and often would not have to bear any consequences in case of a disaster. Reducing costs in most cases reduces the budget for safety measures, but unfortunately higher costs do not mean automatically reduced risks. Costs on one side may increase profit on the other side, without further benefit to the customer. A lot of money is invested in marketing and advertising efforts. This also applies to safety equipment. Suppliers are paid for selling equipment and not for ensuring that it really fulfills its purpose.

While overestimating their own capabilities, professionals tend to underestimate costs and time when preparing budgets. Besides that, costs must be deliberately understated to enable approval of funds. As soon as the project is in progress, the probability that it would be stopped due to increasing costs is low. Such 'salami slicing' is daily routine in any project realization. To save some funds all the same, reserves are reduced, particularly those that are required to handle possible risks.

## 10.10 Professional Ethics

In the good old times, it was the intention of most responsible professionals to do a good job, following superior professional ethics. Lawyers and medics even had to take an oath of sticking to their ethic rules. Engineers were developing good technical solutions, pursuing a properly functioning system and a content client. Content clients supplied you with the next job. Maximizing profit was not an issue, but good work was paid well.

Today, maximization of short-term profits is the primary goal, inevitably encouraging people to take higher risks, deliberately or unconsciously. Rich people are perceived as idols without taking into account professional reputation and where their money really comes from. But there is a significant psychological effect to be considered, as described for instance by Ariely [1]. People like doing good work if they are well reimbursed. Remuneration is not only about the money, but is a way to show appreciation.

There is a limit, beyond which a higher pay is not perceived as an incentive any more. On the other hand, people are willing to work hard without payment when they get the feeling of doing something useful, which is well appreciated in their social setting. According to Ariely, that's when social norms apply, in contrast to the market norm for paid work. The important point is, when people have to work for low remuneration, their commitment decreases significantly. Sloppiness and low quality are the consequences. In a professional context, you often have to do things simply because you are paid for it, even when it may be contradictory to your convictions. For engineers who are mostly led by superiors or clients who lack technical understanding, that is the rule. This aspect confuses and corrupts professional ethics. By having to do and say deliberately what is clearly wrong, intelligent and enthusiastic professionals become blunted cynics. Strictly following orders from remote authorities may accelerate decision making processes, but inevitably leads to conceptual and design flaws and poor implementation. Withdrawing decision-making authority and responsibility from a responsible, assiduous professional will reduce motivation for doing a good job.

The process of public tendering owes a lot to the increase of risks. The more abstract the tendered service is, the less it can be exactly described. Procurers often don't really understand what they are about to tender, particularly in the engineering business. Either the lowest bidder will get the job, or the issuing body will find some legal flaws to assign the contract to his favorites. Under such circumstances, doing a good job would be useless. Even when the client is satisfied, the next job will be tendered anyway. Worst of all, financial pressure leads directly to hidden risks, poor quality, and subsequent additional measures for remediation. Instead of paying engineers, equipment suppliers and workers for doing a good job, a lot of money is wasted for bureaucratic empty runs and legal discussions that actually don't contribute neither to benefit nor to safety. But it employs public servants and lawyers with high wages. Public tendering of works with the intention to save costs finally leads to a significant increase of overall costs and risks.

We live in a world of limited resources, and by wasting funds we miss the chance to benefit from more effective solutions with higher benefit. Funds could be reasonably saved and safety increased by thorough risk assessment and analysis. Mathematically, the cost-benefit ratio of a safety measure can be described as costs per risk reduction, for example the cost per saved human life. Such numbers can be derived from Quantitative Risk Analysis. Saving in analysis prevents finding better solutions.

## 10.11 Challenger Space Shuttle Disaster

Until 1986, the Space Shuttle program, with its reusable, airplane-like space-ship, was acknowledged as the state of the art in space exploration technology, even when it was a disaster from an economical point of view. But then, space Shuttle 'Challenger' exploded soon after the start on January 28, 1986, killing its seven-member crew.

A few moments before the explosion, a flame blowing out of the right solid rocket booster had been visible. In the report on the Space Shuttle Challenger Accident [117], many shortcomings were identified. A ruptured O-ring sealing was identified as the primary source of the disaster. Obviously, the O-rings lost their elasticity in cold temperatures. The Space Shuttle was designed to start in the usually warm weather in Cape Canaveral (Florida), but on the day of the disaster, it was extraordinarily cold, with temperatures actually below the acceptable design conditions. However ruptures had been documented from previous tests. Those and many other technical problems were obviously known to the engineers, but the management ignored the warnings.

Another issue addressed by the investigation report was the chaotic organization and the lack of appropriate communication within the NASA, as well as to its suppliers and subcontractors. The Space Shuttle disaster seems to be a typical example of decision makers ignoring or not even noticing the warnings that were given by the subordinate practitioners. The NASA management was under pressure to launch the Space Shuttle flights, even under unfavorable boundary conditions, fearing that any delays or admitting technical problems would lead to bad publicity. Publicity is essential for an organization that is dependent on public funds. The disaster led to the subsequent cancellation of Space Shuttle flights for more than two and a half years, and the NASA management had to learn a bitter lesson.

Richard P. Feynman (Physicist and Nobel prize winner, 1918 – 1988),
a prominent member of the investigation committee, stated:
'*For a successful technology, reality must take precedence over public relations, for Nature cannot be fooled.*'

# 11 Risk and Safety in Tunnels

## 11.1 Road Traffic

Every day, approximately 3,500 people die in road traffic accidents, mostly young people in developing countries. The risk of road traffic can be derived from the statistical data on road accidents and fatalities, which are publicly available on the Internet in most countries (chapter 14.2). For instance in 2012 there were approximately 28,000 deaths resulting from accidents on European roads and 138,258 in India [88]. One of the reasons for the high accident rate in India seems to be corruption when applying for a driving license [14].

Nevertheless, in most affluent societies, road traffic safety is a success story. Despite the significant increase of traffic volume, there are fewer accidents and fatalities. In case of accidents, fast and effective medical treatment of the injured reduces the number of casualties. For instance in Germany, the ratio of annual road traffic fatalities to the number of registered vehicles has decreased by 94% between 1970 and 2010.

Like with many risks, the human factor is the most important element in road traffic. Humans cause approximately 95% of accidents, and humans suffer the consequences. Only a minor part of the accidents is caused by technical defects and breakdowns. Therefore, the focus of risk mitigation measures should be on drivers' education, improving traffic related decision making abilities and securing visual acuity and reaction speed. Situational awareness of drivers is still a crucial factor. Distraction or drunken driving is penalized in most civilized countries. Driving with sleep deprivation, being a cause for many accidents, is controlled and sanctioned only with professional truck drivers, but obviously difficult to control. Distraction by mobile phones, navigation and audio systems and other modern toys has become a major problem.

In civilized countries, traffic rules and stringent laws, like mandatory safety belts or speed limits, are enforced and encourage safer driving. When experiencing road traffic in developing countries with traditional societies, you become aware that obedience of traffic rules, if there are any, is not a self-evident matter. In societies that have recently developed from traditional cultures to modern civilization, the use of new technologies significantly increases risks. Unlike for traditional risks, appropriate risk awareness has not yet been developed for those new technologies. Thus 90% of the world's road traffic fatal accidents occur in developing countries with less than 50% of the world's registered vehicles.

The most important factor influencing traffic risk is driving speed. Kinetic energy is proportional to the square of speed, and energy means potential damage. Due to fixed reaction times and extended braking distance, the probability of an accident increases with speed too. We know that driving speed should preferably be adapted to the circumstances and boundary conditions, like characteristics of the road, driving surface, sight distance, traffic conditions and weather. But most of us tend to drive faster than we can afford, taking the risk deliberately. Feelings like the thrill of speed or time pressure are stronger than rational thinking. Speed reduction is the most effective and economic form to increase road safety, but not very popular. In civilized countries, road authorities have issued speed limits, although it would be more effective to limit the speed of vehicles by technical measures. The freedom to rush, which is a strong emotional sales argument of the powerful automotive industry, is obviously given a higher importance than traffic safety.

Fig. 25    Enforced speed limit: Most efficient safety measure *(Public domain)*

Modern vehicles have a higher efficiency and are much cleaner and safer than their predecessors. Passive safety features like reinforced passenger compartments and impact zones, and active equipment, for instance Airbags, ABS (Antilock Braking System), ESP (Electronic Stability Control) and others, are the standard in new cars. Safety equipment can possibly increase risky behavior of drivers through the psychological effect of risk compensation, but in general, its advantages prevail.

Actually, there is an ongoing technical development towards driverless, automatically guided road vehicles. Like all automation, that would undoubtedly increase safety. On the other side, people would lose the proficiency in driving by themselves, their awareness and their personal responsibility.

## 11.2   Tunnel Safety

Tunnels represent the most expensive type of transportation infrastructure, since they are more elaborate to construct than open roads, bridges and other civil engineering structures. We would not build them if they had no substantial benefits. Therefore, when addressing the issues of risks in tunnels, it must be pointed out that a tunnel itself must be understood as a risk mitigation measure. In comparison to an open road, tunnels usually provide a faster and safer means of transportation, shortcutting dangerous mountain or city roads, protecting the road users from external hazards like bad weather, rockfall and avalanches, and in return shielding the environment from the negative impact of road traffic, like noise and pollution. Such basic benefits are often neglected when talking about tunnel safety.

Fig. 26    Curve in a road tunnel  *(Public domain)*

Generally, there are fewer accidents in tunnels than on open roads, but accidents are comparably more severe. Furthermore, the collision risk is much lower in a highway tunnel with unidirectional traffic, than in a tunnel with bidirectional traffic. The fear of tunnels mainly seems to be a psychological problem with an element of claustrophobia. An important issue are curves in tunnels, where the sight distance is reduced in comparison with an open road. As a consequence, the design speed must be adapted to the stopping sight distance, which is defined by the geometrical conditions. Another aspect is limited access, which is usually only possible through the tunnel itself or a parallel tunnel tube or separate service or emergency access tunnels, if there are any. That may impose a restriction to self-rescue and operations of emergency services in case of an incident in the tunnel.

There is a significant difference between open roads and tunnels when a fire occurs. Therefore, when talking about road tunnel safety, the issue is mostly about fire safety. Unlike accidents, many fires are not caused by human faults, but by technical failures. Fires happen regularly in road tunnels, but mostly don't lead to any casualties and only cause minor damage. I worked only on two road tunnels where fires used to occur in statistically significant numbers: In the 17 km long Gotthard road tunnel in Switzerland, there had been an average of approximately 4 fires each year. However, until 2001, there hadn't been any fatalities due to fires. Since then, the accident and fire rates have been reduced mainly by traffic restrictions. In the four (previously three) 3 km long tunnel tubes below the Elbe river in Hamburg (Germany), one of the most frequented tunnels in Europe, on average approximately 10 incidents with smoke in the tunnel happen yearly. Only few of them are really dangerous fires, and they haven't caused any fatalities so far, possibly thanks to extensive safety measures, for instance stand-by fire brigade on site.

Before the fire incident in the Mont-Blanc tunnel 1999 with 39 casualties, less than 100 people had died in history due to fires in road tunnels worldwide in peace times[30]. Such a number is absolutely negligible, taking into account that road traffic accidents claim nearly 3,500 lives daily worldwide. As a rough estimation, only one of 1 million road traffic fatalities dies in a tunnel fire. Thus fires in road tunnels seem to be a typical example of risks that are overestimated in public opinion by extensive media coverage. Fires in mass transport facilities like metro and rail tunnels happen less frequently, but may lead to much higher numbers of casualties.

---

[30] War incidents, like the one described in chapter 3.6, are not taken into account in that figure.

Since the probability to die due to a fire in a tunnel is negligible, any safety measure might be questioned from the point of view of appropriate allocation of limited funds. But if you find yourself in a tunnel filled with toxic smoke, without means to escape, statistics are only of small consolation.

The risk in tunnels is dependent on many factors, like traffic volume and composition, speed, geometrical conditions and tunnel equipment, and most importantly, the drivers' behavior. In fire safety issues, it is mostly the smoke that kills people, and inside a tunnel, smoke can spread very quickly. In contrast, the heat can cause massive material damage. Important secondary safety measures are about providing means of escape to affected people and smoke control. After a fire in a tunnel, even without direct casualties, the tunnel needs often to be closed for repair. In the meantime, traffic might be directed over dangerous bypass roads. When severe accidents happen on those roads, is the resulting damage to be considered a consequence of the original tunnel incident?

Tunnel fire safety can be seen in principle as quite a simple issue. Even when going into the details makes things more complicated, it is important to remember the basic principles. In fire incidents, it is mainly the smoke, respectively its toxic components, that endangers people. The linear structure of tunnels may lead to a fast linear smoke spread. When the smoke moves faster than people can escape, the situation gets very dangerous. Fast smoke movement kills people! And yet, a high airflow velocity, leading to fast smoke movement in case of fire, is the design case according to most tunnel ventilation standards and guidelines.

Without forced ventilation, the smoke moves predominantly by natural forces, which cannot be controlled, and by traffic as long as it keeps moving. In steep tunnels, the chimney effect by the fire can get very strong and drives the smoke quickly upwards. This caused for instance 156 fatalities in the Kaprun funicular train in Austria in 2000, or 9 fatalities in the Viamala road tunnel in Switzerland in 2006. But fast smoke spread can also be caused deliberately by the operation of the ventilation system. Fire ventilation is often understood as simply driving the smoke away from the fire location. People on one side of the fire get the chance to escape in a smoke-free area and emergency services get better access to the incident site. This works fine as long as there are no people exposed to the smoke on the other side, where it is blown, for instance in highway tunnels with unidirectional traffic and low probability of congestions.

Unfortunately many operators, designers and 'experts' don't really understand the difference between design and operation. In practice, pushing the smoke towards one side is often performed without considering where endangered people may be. That cost approximately 300 fatalities in the metro Baku (Azerbaijan) in 1995, and also in the Mont-Blanc, Tauern and Gotthard road tunnel fires as described herein.

The actual state of the art of fire ventilation in tunnels is the closed loop control of longitudinal airflow, based on quick fire detection, precise and reliable, multiple measurements of the air speed, and continuous control of fans by frequency converters. By that, smoke movement can be stabilized to a reduced velocity that is slower than the supposed escape speed of people, and in the vicinity of the fire a smoke stratification is facilitated, providing tenable conditions for the escape of people. The closed loop control of longitudinal airflow was developed after the year 2000, mostly in Swiss tunnels, and meanwhile has reached a high level of reliability. Details are described in the Tunnel Ventilation Compendium [109].

Fig. 27    Stabilization of smoke stratification, providing safe escape conditions

## 11.3   Personal Approach

Prior to working on road tunnel projects, I had gained professional engineering experience in building technologies like heating, ventilation and air conditioning, and industrial plant design. Responsibility for the design, implementation and proper function of a system was a matter of course. When taking over some tunnel ventilation projects, I got the opportunity to work with the leading specialists in the field. Amongst other activities, my job was the supervision of the installation, commissioning, acceptance and initial operation of ventilation systems in tunnels prior to opening.

In such phases of system implementation and commissioning many things go wrong, problems have to be fixed, the deadlines are short and the nerves of all involved professionals are on edge. According to previous professional experience, I didn't expect anything else than the fulfillment of Murphy's law, but the dilettante approach of many involved professionals exceeded my expectations. Immediately after the opening of a tunnel where I had worked on the commissioning of the ventilation system, road users complained about poor air quality in the tunnel. Moreover, fogging of windshields of cars entering the tunnel caused accidents[31]. After some careful consideration and discussion with my colleagues, I found out that the problems were inherent in the ventilation concept, which was not useful in principle. Details are explained in the Tunnel Ventilation Compendium [109]. That fact had been even mentioned in earlier publications. Practitioners confirmed that there had always been problems, but they simply did what they were told without questioning. Peer pressure and group consensus led responsible professionals to design and apply inappropriate ventilation systems despite actually knowing that they were doing the wrong thing.

In another tunnel where I was about to work, elaborate airflow measurements for acceptance tests of the ventilation system were worked out. I wanted to compare the results with the data from the installed flow measurement. The measured values were completely wrong, but nobody cared about it. It hadn't occurred to the designer of the measurement systems to even prove whether the measurement provided the right data. They simply relied on it, without questioning. Others complained that the flow measurement was unusable, but did nothing about it. What was the measurement for? Obviously it had no purpose, besides indicating some (wrong) values on a computer screen that nobody cared about.

This changed radically when after 1999, the control of longitudinal airflow became an important issue. To achieve that, reliable and precise flow measurement was essential. Subsequently, the designers and equipment suppliers invented detailed procedures for the conception, implementation, calibration and signal analysis for such measurements. Until the 1990s, tunnel ventilation systems had mainly been designed to reduce the impact of noxious gases and reduced visibility in the tunnel under traffic conditions. After the introduction of vehicle emission restriction laws in most western countries, that purpose became less important.

---

[31] Collisions due to fogging of windshields are an underestimated risk in many road tunnels in cold, humid climate.

Fire ventilation had only been a minor issue, until the disastrous fire incidents in 1999 and the subsequent media attention. Fire and smoke tests had been carried out regularly, and even some excessive investigation programs were worked out. Those tests showed that most ventilation systems were not really useful in controlling the spread of smoke. Many tunnels ended up filled completely with smoke, and it took a long time to remove it. In most fire and smoke tests, the ventilation system obviously did not fulfill its purpose, but nobody seemed to care. All involved professionals, including the client's representatives, simply accepted that it didn't work, without questioning it. In fact, in those times, almost no tunnel ventilation system really worked properly. There were no clearly defined goals to be achieved, and there were no responsibilities assigned. The ventilation designers cared about the fan layout, and the electrical designers, who did not really understand ventilation issues, were responsible for the system implementation. Many 'experts' did not (and still don't) understand that what really counted was the control and operation of the whole system in practice rather than the design on paper. However, the final goal is not the proper function of a technical system, but to protect and save the life and health of people, and to prevent damage to the infrastructure.

Only after the fire incidents in 1999 and 51 fatalities, the road authorities in European countries finally changed their attitude. Since then, an advance was made in fire ventilation issues, but the dilettante approach is established too well within the system. In some aspects the conditions have even got worse. Strict application of guidelines without scrutinizing and taking into account cost benefit considerations often leads to absurd conceptual decisions. Technical understanding is replaced by bureaucratic procedures.

Once when I worked on a tunnel with an exhaust ventilation system, the quality of construction works had been an issue, as in most projects. Holes in the concrete structure of an air duct on the pressure side of exhaust fans would lead to contaminated exhaust air and in case of fire smoke entering the whole ventilation station. That would affect the electrical equipment and possible occupants. In fact, such is regularly observed during smoke tests and real fire incidents. I addressed the issue many times in the regular meetings of the project team. The deficiency was noted again and again in the meeting reports, among many others, but unfortunately the task got forgotten. In fact, nobody from the supervisors seemed to care about how the site really looked, because most of the team was completely absorbed by bureaucratic paperwork. After some months and approximately a dozen meetings, I just grabbed a trowel and a bucket of mortar, and closed the holes by myself. Not surprisingly, the then project manager is today a senior official in a road authority.

## 11.4   Tunnel Fire Disasters

The fire incidents in the Mont-Blanc tunnel (France/Italy) and Tauern tunnel (Austria) in 1999 as well as in the Gotthard Road tunnel (Switzerland) in 2001, led to extensive public media attention and discussions about tunnel safety. The media impact of the incident in the Viamala tunnel (Switzerland) in 2006 was lower. By then, the hype had faded. I had worked on the Mont Blanc tunnel, the Gotthard Road tunnel and the Viamala tunnel projects, therefore the following information is authentic, mostly provided by my own first-hand experience, people I personally known, and the PIARC publication [101].

A key role in all those incidents played the smoke propagation and tunnel ventilation as well as the availability of safe shelters and escape-ways, demonstrating the usefulness and limits of safety measures. All victims suffocated from toxic gas, except the people who had been killed by the preceding accidents, mainly in the Tauern and Viamala tunnels. All four tunnels had one tube with bi-directional traffic, therefore there were vehicles and people on both sides of the fire location. Only the Gotthard tunnel had emergency exits through cross passages to a parallel escape tunnel. In the Mont-Blanc, Tauern and Gotthard fire incidents, the tunnel ventilation was not only unable to control the spread of smoke in the long tunnels, but even accelerated the smoke propagation. The short, steep Viamala tunnel was quickly filled with smoke due to the chimney effect, and the ventilation system was kept switched off, since it would have been useless anyway.

### Mont-Blanc 1999

The Mont-Blanc tunnel is situated below the highest peak of the Alps between France and Italy. When opening in 1965, it was the longest road tunnel in the world with 11.6 km length. The tunnel was operated by two independent companies, one French and one Italian, from two distinct control centers. Communication and coordination between the two operating companies, in two different languages, was quite poor. On March 24, 1999 a fire started inside the tunnel, originating from the engine of a Belgian truck loaded with margarine and spreading over various other vehicles. Totally 34 vehicles including 20 trucks burned. According to the official report [37], the fire was caused most probably by a cigarette stub in the air filter of the truck engine. The fire site, where the truck had stopped, was situated near the center of the tunnel. Fire detection was delayed, and the number of cars in the tunnel was unknown to the operators. Traffic signals, preventing vehicles from advancing further into the tunnel, were either not obeyed or not working at all.

The smoke extraction capacity of the tunnel ventilation system was underde-signed, like in all tunnels at that time. After the fire alarm, the Italian operators blew fresh air into the eastern side of the tunnel, thus providing a safe zone for the people in the Italian part, but accelerating the smoke all over the western (French) part. There, 27 tunnel users died resting in their vehicles, and 10 died while escaping on foot. Two rescue workers lost their life too. However, many people have saved their lives despite the unfavorable conditions, because they immediately left their vehicles and fled on foot to the remote portals, as did the driver of the truck that started the disaster.

Fig. 28    Inside the Mt. Blanc tunnel after the 1999 fire  *(Public domain)*

An important aspect is that the Mont-Blanc tunnel was equipped with shelters every 600 m, supplied with fresh air, but without access to the open. In a shel-ter close to the fire zone, two people died. The shelters had a two hour fire protection rating, while the fire exceeded temperatures of 1000°C and lasted for 53 hours. It took five days for the fire site to cool down sufficiently to enable unprotected access. As a consequence, subsequent new tunnel safety rules banned closed shelters, demanding escape ways to the open. Nevertheless, some tunnel users and fire fighters have survived the Mont-Blanc tunnel fire only thanks to those shelters, where they were protected from the deadly fumes, until they were rescued by the fire brigade. Also, the importance of a safe drainage system was derived, since the spreading of burning liquids over the road surface had caused further fire propagation.

## Tauern Tunnel 1999

Only two months after the Mont Blanc tunnel fire disaster, 12 people died on May 29, 1999 in a rear-end collision and a subsequent fire in the 6.4 km long Tauern tunnel in Austria. A sleep deprived truck driver crashed into a column of several vehicles, including a heavy goods vehicle with lacquer tins. That column had built up in front of a stop light at a construction site inside the tunnel. eight fatalities resulted from that accident, regardless of the following fire, where 20 cars and 14 trucks burned down during 14 hours. Temperatures in the tunnel exceeded 1000°C. Three victims were found dead in a vehicle that was 100 m away from the fire site towards the portal, whereas people that were closer to the fire had successfully escaped. The last victim was found 800 m from the fire towards the interior of the tunnel. He was suffocated when coming out of an emergency call niche. Three people took refuge in another emergency call niche and were saved later by the fire rescue service.

Fig. 29    Fire in the Tauern tunnel 1999  *(Public domain)*

The people in the Tauern tunnel had been sensitized to tunnel fires, because the Mont-Blanc disaster was present in all media by then. Approximately 80 people managed to escape from the tunnel. The ventilation had pushed the smoke more than one kilometer towards the inside of the tunnel. Later, the flow direction was changed on demand of the fire brigade, in order to save the people in the emergency niche.

## Gotthard Road Tunnel 2001

The 17 km long Gotthard road tunnel in the Swiss Alps was the world's longest road tunnel at that time, being situated on an important European North-South traffic link. On October 24, 2001, a frontal impact between two trucks occurred, caused by a Turkish driver in a Belgian truck, resulting in a fire in which 10 cars and 23 trucks were totally burned or damaged. The fire spread so quickly that it could not have been fought by hand held fire extinguishers. Burning tires emitted dense smoke. The driver that caused the crash was found dead 300 m away from his vehicle. According to the investigations, he had been drunk.

Fig. 30    Fire in the Gotthard road tunnel 2001    *(Public domain)*

When the accident occurred, the refurbishment of the tunnel ventilation system towards a controllable concentrated smoke extraction had already been in progress. Unfortunately, the old system with line exhaust was still in operation when the incident happened. The fire detection system gave further alarms, which activated the linear smoke extraction system in other sections, increasing the air velocity at the fire site and further fanning the flames and pulling the smoke more than 3 km towards the inside of the tunnel [32].

---

[32] A basic rule in tunnel fire safety issues is that only the first confirmed alarm must release the automatic response. Working on the ventilation design, I pointed that out in a technical report several weeks before the fire incident happened. If that were applied, smoke propagation would have been significantly reduced.

The Gotthard road tunnel is equipped with emergency exits to shelters in distances of 250 m, which are connected to a parallel escape tunnel. Those escape facilities saved at least 30 tunnel users. Others escaped towards the south portal. Most of the 11 people who died could have survived if they had reacted quickly and properly, fleeing immediately to the shelters. Those that did not were trapped in the smoke, giving an example of irrational fatal human behavior. Five victims died staying in their vehicles. One driver was found next to an emergency exit.

## Viamala 2006

The situation in the 750 m long Viamala tunnel was different from the previously discussed fire incidents. The old, steep and narrow tunnel has tight curves, where the sight distance is limited. On September 16, 2006, an accident between a bus and two cars resulted in the outbreak of a fire near the lower end of the tunnel. The fire grew quickly, and the chimney effect caused uncontrolled spread of smoke and toxic gases to the upper portal. Seven vehicles were trapped in total. Subsequently, the tunnel was filled with smoke over the whole length very quickly, before any safety system could react. The jet fans of the ventilation system stayed switched off, since they were of no use to control the smoke spread effectively.

Fig. 31    Burnt car and bus in the Viamala tunnel 2006  *(Public domain)*

Nine people died, some due to the accident, others were poisoned by the fumes. A family of four died in their car, another four people were found dead on the road surface in the tunnel, and one truck driver died later in the hospital from his injuries. However, 21 tunnel users, including an ice hockey team, were able to escape.

When the fire fighters arrived on site, the tunnel equipment in the vicinity of the fire site had already been destroyed and fallen to the surface. Downstream of the fire, they had to search for people under zero visibility conditions. Finally the fire could be extinguished and the smoke was removed using a mobile jet fan.

The Viamala tunnel had been identified as dangerous before, but planned safety measures had not been implemented yet by then. Until today, the Swiss road authorities did not address the primary cause of the accident, which is the insufficient sight distance at the allowed speed in the narrow tunnel. Ten years later, a ventilation refurbishment program with fast smoke detection and controlled ventilation with more powerful jet fans has been successfully completed and a parallel escape tunnel was under construction. However, the cost efficiency regarding the risk reduction by this escape tunnel is another issue.

While I was writing this book, a fire occurred in the escape tunnel construction site, and the smoke was sucked to the main tunnel through partially open cross passages. Almost the whole tunnel was filled with smoke, impairing traffic. One driver suffered from light smoke poisoning. All safety systems worked well, but unfortunately the mode of operation was not useful for this particular situation. It was an event that nobody has thought about previously.

### Résumé

As a result from the four described tunnel fire incidents, totally 71 people lost their lives, and subsequently billions of Euros had been invested for refurbishments and new safety measures in many old and all new European tunnels[33]. The tunnels were closed for months or even years. This led to traffic diversion to less suitable, more dangerous routes. Beside the ecological impact of additional kilometers driven, the increase of accident rates on the alternative routes led to additional casualties as a secondary effect. However, the magnitudes of the mentioned incidents were on different levels. The Mont-Blanc tunnel fire can be seen as an extraordinary catastrophe, exceeding by far the average fatality fee of road traffic incidents, leading to a subsequent three-year closure of the tunnel. The other described road tunnel fires were not extraordinary in comparison with other serious traffic accidents and tunnel fires with similar damage that had happened before. The Tauern tunnel was cleaned and repaired within three months, the Gotthard tunnel within two.

---

[33] 380 Mio. EUR for the refurbishment of the Mont Blanc tunnel alone.

Most casualties in the Tauern and Viamala tunnels died in the preceding traffic accidents, not due to the fires. Most of the victims of the Gotthard road tunnel fire incident, where emergency exits to a parallel escape tunnel were situated, died because of their inappropriate reaction, having the possibility to escape. The basic lessons from the 1999 fire incidents were the importance to provide escape ways and ventilation systems that are able to control the spread of smoke in the tunnel, preferably with concentrated smoke extraction, as is described in the Tunnel Ventilation Compendium [109]. However, the lessons from the mentioned tunnel fires apply mostly to long single-tube tunnels with bidirectional traffic. For highway tunnels with unidirectional traffic, the situation is completely different, but that was overlooked in the subsequent elaboration of tunnel safety standards.

## Amendment – Yanhou Tunnel 2014

While I was about to write this book, one of the worst road tunnel fire disasters was reported in northern China on March 1 2014. Unlike the other described examples, where I have obtained personal, first and second-hand information, the following information is taken from Internet sources.

Two methanol tanker trucks crashed during traffic congestion inside a short twin bore highway tunnel on the Erguang Expressway. The drivers tried to examine and disentangle the trucks, triggering a fire that quickly spread to other trucks, carrying coal and other flammable materials. About 100 Minutes after the crash, a liquid natural gas tanker exploded. It took 73 hours to extinguish the fire.

Fig. 32    Smoke exiting the portal of the Yanhou tunnel *(Public domain)*

In the Yanhou Tunnel, safety equipment did either not exist or not work. The cross passages could not be used for escape purposes. Many important safety measures seem to have been neglected. A queue of coal trucks built up inside the tunnel, and even after the outbreak of the fire, the tunnel was neither closed to traffic nor evacuated. At least 31 people died, 9 were missing and 42 vehicles were destroyed. But even then, the death toll was less than 5% of the average daily number of fatalities on China's roads.

## 11.5  Smoke Test

The Branisko tunnel is situated in Eastern Slovakia and consists of a 5 km long tube with bidirectional traffic and an escape tunnel parallel to the main road tunnel. It serves to bypass a dangerous mountain road where fatal accidents used to happen regularly. I was responsible for the operation and control algorithms, the commissioning and the acceptance tests of the tunnel ventilation system, which by the way had been designed not really adequately by another company. Beside that, I worked out the concept, design and realization of the escape tunnel ventilation system, which later served as the prime model for all Swiss escape tunnels.

As a final acceptance test before the tunnel opening in 2003, smoke tests were carried out to demonstrate the effectiveness of the fire ventilation system. It has to be pointed out that on beforehand the function of all safety systems should have been successfully proven. To simulate tunnel fires, we proposed to use military smoke grenades that were designed originally for the screening of battle tanks, producing a big amount of artificial smoke. Health aspects were not addressed then, since the test smoke was quite irritating to the respiratory system. The first test was a success. Although the smoke extraction capacity of the ventilation system was not sufficient, the fire ventilation system was able to limit the spread of smoke and finally to remove the fumes from the tunnel.

Unfortunately, during the second smoke test, a breakdown of the power supply occurred. This led to the failure of an exhaust fan, and additionally of one supply fan in the opposite tunnel section which was used to slow down the longitudinal airflow. As a consequence, the smoke spread over the whole length of the tunnel due to strong buoyancy forces. I attended the smoke tests in the tunnel standing beside a cross passage to the escape tunnel, and left my car in a lay-by opposite the cross passage. When the tunnel filled up with smoke due to the system failure, I decided to evacuate myself to the cross passage to the escape tunnel, which was supplied with fresh air.

After a few minutes, I realized that my dog was trapped in the car, and that the smoke could affect his health, so I decided to pick him up. The distance between the cross passage and the car was approximately 20 meters. While walking in the tunnel, I was holding my breath to avoid respiring the fumes. I found my car quickly, took out my panting dog, and tried to get back to the cross passage. However, it took longer than expected to reach the opposite tunnel wall, which was only a few meters away. The visibility in the smoke was practically zero. While trying to find my way along the tunnel wall to the cross passage, I suddenly realized that I might be going in the wrong direction. Meanwhile, I could no longer hold my breath and had to respire, using my handkerchief as an improvised filter.

Fig. 33    Smoke test in the Branisko tunnel

Following along the wall in the opposite direction, I finally reached the fresh air cushion which built up in front of the cross passage. That was an effect of the air streaming through the pressure relief damper, which was situated over the escape door. Happy, but coughing, my dog and I entered the safe space. This was a simulated emergency when I knew that my life was not threatened and where I had a good overview of the situation. Despite knowing the positions of the lay-by and the cross passage, I had problems finding my way through the smoke along a distance of only a few dozen meters.

This little personal adventure showed impressively that people may fail to reach escape ways in a tunnel filled with smoke and zero visibility even on short distances. Most casualties of real fire incidents die due to smoke exposure while losing orientation. Many car drivers don't even notice the position of the escape doors. Another alarming aspect was the reliability of safety equipment, which was supposed to work reliably, having been recently tested. I had personally attended other smoke tests in tunnels in different countries, where the fire ventilation failed due to a breakdown of the electric power supply, when all systems were supposed to have been thoroughly tested out previously. There are plenty of examples of such system failures in practice, but operators and responsible designers are not keen on publishing them. In real incidents, such failures may cost human lives.

## 11.6  Fire Safety Measures

The findings from the road tunnel fire incidents, mainly the 1999 Mont Blanc disaster, were not really new. In fact, many tunnel safety and ventilation practitioners had been aware that for instance the ventilation systems were unable to confine smoke spread. However, they were not heard. Incidents happened regularly, but with low frequency, and there were very rarely any casualties. Many road authorities reduced budgets for safety measures.

Extensive media attention and subsequent political pressure after 1999 led to new national and international tunnel safety rules [33], focusing on:

- Providing twin-tube tunnels with unidirectional traffic
  when the traffic load exceeds a certain threshold level
- Escape ways from tunnels to the outside in minimal distances –
  closed rescue shelters are not allowed
- Fast, reliable incident and fire/smoke detection
- Efficient traffic management systems in- and outside tunnels
- Control of smoke propagation by control of longitudinal airflow and,
  in long tunnels with bidirectional traffic, concentrated extraction
- Sewage systems for containment of spilled flammable liquids
- other constructional measures, for instance lay-bys
- Risk analysis as basis for decisions
- Organizational and operational measures
- Education of drivers, operators and emergency services

Those measures increased road tunnel safety significantly, and allowed for large funds to be invested, providing good business opportunities for the construction industry and equipment suppliers.

Good ideas and technical concepts of the guidelines were never really scrutinized regarding their practical implementation. Ongoing advancement and practical experience from the application of the guidelines have been consequently neglected and even suppressed by the authorities. Therefore, the new standards that defined requirements to safety equipment were soon overtaken by the reality of professional and political dilettantism. The huge costs also led to the delay of tunnel projects, in some cases even to their postponement or cancellation. In contrast, proposals for quick and simple measures to increase safety of existing tunnels were deliberately ignored by the road authorities. That way, the road safety was in fact reduced, when taking into account the lengthy unsafe states on the long run.

Most proposed safety measures in road tunnels, like escape ways or ventilation systems, are in fact secondary measures to mitigate the damage in case of fire. Since the probability of such incidents is very low, any such measure might be questioned from the point of view of a cost benefit analysis. Primary measures, reducing the probability of accidents, would be more important. Like on open roads, this is mainly influenced by the driver's behavior. Key elements are alertness, visibility, speed and distance between vehicles. The most effective and cheapest measure would be enforcing a lower speed limit, however such a measure is often politically not feasible. At least, the allowed speed should be adapted to match visibility conditions, particularly in curves, but even that is not guaranteed.

**Twin-tube Highway Tunnels**

The most expensive, but efficient risk mitigation measure in a single tube tunnel with bidirectional traffic is to separate traffic directions by providing a second tunnel tube. In fact, twin-tube highway tunnels with free flowing unidirectional traffic provide one of the safest types of roads – if there are no perpendicular walls. You pay almost twice the price of a single tube, but you reduce the risk significantly. Even in case of fire, the risk in such tunnels is very low, provided that the vehicles can safely leave the tunnel between the incident site and the exit portal. Blockage of vehicles can and must be avoided by appropriate traffic management measures, like closing the tunnel entrance portal if necessary to avoid congestions inside any tunnel

Irrespective of the tunnel length, the smoke can be blown in the direction of traffic movement, since vehicle speed is always faster than the speed of smoke propagation. When the flow velocity exceeds a threshold value, backlayering of smoke will be avoided. People who are stuck in front of the incident site are situated in a smoke-free, safe area, see Fig. 34.

Fig. 34    Fire ventilation and direction of escape in a tunnel with unidirectional traffic

An often-discussed issue is smoke extraction in tunnels with unidirectional traffic. This may be useful when vehicles and passengers get trapped downstream of the incident site, for instance in case of traffic congestion in an urban tunnel. However, in twin-tube highway tunnels, the passengers still have a means of escape over the cross passages to the other tube. There might be other arguments to be taken into account, for instance the benefit of a smoke extraction for occasional situations with bidirectional traffic in one tube.

In contrast, the smoke extraction may also lead to additional risks. Exhaust systems in road tunnels are mostly realized by a slab that separates the traffic space from the air duct in the ceiling. There were incidents when this slab had fallen down and killed people. This happened for instance in 2006 in the Boston Central Artery Tunnel (U.S.) where a woman was killed, and in 2012 in the Sasago Tunnel (Japan) with 9 people killed.

Fig. 35    Collapsed slab in the Sasago tunnel 2012   *(Youtube /Public domain)*

In a Swiss highway tunnel, the slab had fallen down too, but luckily nobody was harmed. While I was writing this book, in an Austrian tunnel the recently constructed slab subsided. There are probably many such incidents, but usually kept secret.

Another situation, in which smoke extraction leads to an additional risk, is a possible erroneous operation of the extraction. When opening the exhaust dampers upstream of the incident site, for instance due to false detection, the area where blocked cars and people are located would be smoked-up. This happens regularly in real fires and smoke tests. In some cases, it was even applied deliberately by designers or operators who do not really understand the fire ventilation principles in a road tunnel.

Fig. 36    Wrong operation of extraction in a tunnel with unidirectional traffic

To sum up, there are situations when the extraction in a highway tunnel would have a possible benefit, and other cases when it increases the risk. Most of the time it would have no effect on safety at all. However, smoke extraction significantly increases costs, usually by dozens of millions of Euros in one tunnel, mainly for constructional provisions. Summarizing, smoke extractions for highway tunnels with unidirectional traffic have a limited usefulness, increase the risk in particular aspects, and cost a huge amount of taxpayer's or driver's money. Despite that fact, design guidelines in some European countries demand smoke extraction for long highway tunnels with unidirectional traffic, giving a typical example for imprudent requirements.

**Escape Ways and Shelters**

Whenever you drive through a road tunnel, you should notice the distance between emergency exits – if there are any. Emergency exits from tunnels are an important measure to improve fire life safety, but the usefulness of escape ways from tunnels is a complex matter, influenced by factors deriving from fire scenarios, spread of smoke and escape of affected people. Escape ways mitigate the consequences of damages as a secondary measure, but have no influence on the probability of occurrence of incidents. Therefore, their effectiveness in saving tunnel users lives is limited in principle, and depends strongly on the boundary conditions.

Fig. 37     Emergency exit from a road tunnel

As a consequence of the Mont Blanc tunnel fire disaster, escape ways from tunnels to the open were recognized as important safety measures. For instance, the European Directive [33] requires escape ways from all tunnels on the Trans European Road Network to be within a distance of at most 500 m. Many national guidelines define distances between 200 m and 300 m. But those distances were not fixed on any cost benefit considerations, but rather on simple assumptions taking into account existing solutions[34]. I doubt whether any of the authors of those guidelines has ever tried to escape from a tunnel filled with smoke under zero visibility conditions. As simulations have shown, as well as my own personal experience in the Branisko tunnel, the distance of emergency exits should not exceed approximately 100 m to be effective for saving people in a tunnel fire incident. For longer distances, the usefulness deteriorates disproportionally.

---

[34] For instance, the distance of emergency exits in the Gotthard road tunnel was fixed at 250 m in the 1970ies, foreseen as cross-passages to a future second tunnel tube. This distance was adapted later as default for many other tunnels.

On the other side, when the ventilation system can guarantee an effective, reliable control of smoke movement, escape through the traffic space would be possible over long distances. Any equipment like ventilation systems can fail, for instance in case of a power supply failure. In contrast, escape ways are always available, as long as the doors can be opened. Additional technical measures, like locks and overpressure, prevent smoke from entering the escape way. Unfortunately, the overpressure ventilation that keeps escape ways smoke free in case of open doors may generate an excessive pressure on closed doors, preventing their opening. Technically, this problem can be handled, but a useful solution is often impeded by inapt design guidelines.

However, when taking into account cost benefit considerations, the general demand for escape ways from tunnels according to present standards might be questioned in principle. In fact, emergency exits are a 'completely or not at all' measure. Providing emergency exits within short distances from single tube tunnels requires the construction of additional escape tunnels, either parallel or perpendicular to the main tunnel at an enormous cost. A Quantitative Risk Analysis for such an 'average' single tube tunnel has shown that the cost per life saved of such a measure exceeds one billion EUR!

In contrast, where two tunnel tubes are separated only by a dividing wall, for instance in cut and cover tunnels, connections between the tubes can be realized very simply. Escape ways from one tunnel tube to the other in short distances are available almost 'for free'. Nevertheless, in many cases narrow-minded executives required even in city tunnels with a single dividing wall to strictly apply the maximum distance between emergency exits according to the guidelines, deliberately impairing possibilities to escape and missing a chance for significant enhancement of safety at negligible costs.

The ban of closed shelters is a direct consequence of the Mont-Blanc tunnel fire in 1999. In fact, two people (of totally 39 fatalities) died in a closed shelter in the Mont-Blanc tunnel, but it is neglected that the other shelters had in fact saved many tunnel users and firefighters. Two months later in the Tauern tunnel, three people have survived in an emergency call niche.

Most fatalities in tunnel fires die of suffocation or poisoning from smoke inhalation. Closed shelters, supplied with fresh air, could save such people at significantly lower costs than parallel escape tunnels. Such shelters are provided as standard in many subterranean construction sites and mines, where the risk is usually higher than in road tunnels. A residual risk of fatalities being trapped in shelters in the immediate vicinity of a disastrous fire, as it happened in the Mont Blanc tunnel, must be accepted.

Fig. 38    Closed shelter in an underground construction site

## Influencing Human Behavior

As in all risk issues, human behavior is the key for tunnel safety, but strongly neglected by many engineers who focus exclusively on technical systems. Measures for general traffic accident prevention, like alertness, reduced speed and distance between vehicles apply in tunnels as well. In case of fire, an immediate escape would be essential. Most casualties of tunnel fire incidents refuse to escape, preferring not to give up their vehicles, where they feel safe. Therefore, it would be highly important to educate drivers to leave their cars and move away from the smoke to the emergency exits right away when alarmed or when they see smoke in the tunnel. For being able to do so, tunnel users must notice where the emergency exits are before a fire would break out. Think about that when you drive through a tunnel the next time.

Videos from surveillance cameras sometimes show people fleeing right into the smoke. In most tunnels, the distance to the nearest emergency exits is indicated by standardized signals, similar as in buildings. It would be more useful to indicate that direction of escape contrary to the smoke movement. In highway tunnels with unidirectional traffic, that would always be against the traffic direction, without considering the distance to the nearest exit (see Fig. 34). Unfortunately, escape signage in most highway tunnels is contradictory to that.

Fig. 39    Misleading escape signage in a highway tunnel

## Incident Detection

Active equipment, for instance the alarming of drivers, stopping signals and emergency ventilation and lighting, must be switched on in case of an incident. To achieve that, a fast and reliable incident detection is necessary. Unfortunately, sensitivity and reliability are contradictory requirements. The fastest detection systems like CCTV signal analysis imply a high false alarm rate, and therefore they cannot be used in practice for automatic release of safety systems. In contrast, fire detection systems that are based on temperature measurement are generally too slow in tunnel applications and often not able to detect a fire at all, for instance while a fire is smoldering inside a vehicle. In practice, smoke detectors, based on simplified visibility measurement, have been proven as most reliable and fast fire detection systems in tunnels. A tunnel with poor visibility may constitute a source of accidents and should be closed for traffic, without the need to consider whether it originates from a fire or from dust dispersal. On the other hand, smoke detectors cannot reliably determine the exact location of the fire, which might be required for smoke extraction, traffic management and other safety systems.

## 11.7 Adjusting Standards and New Guidelines

In a 'not to be named', mountainous country, a big engineering company was responsible for the design of the ventilation system of a new tunnel. At that time, in the mid-1990ies, continuous extraction systems were the rule, and the extraction capacity was defined by design standards. During commissioning, tests on the ventilation system showed severe underperformance. The system fell short of the required extraction capacity by approximately 25%. Errors happen, and usually have to be fixed. But in that case, the responsible civil servant from the public road authority[35] decided that the ventilation was sufficient. Instead of increasing the ventilation capacity to achieve the required value, the standard was simply adjusted. The new design rule reduced the previously required value by 25%. The design team and the public authorities were proud that they had met the new standard and even reduced costs.

In the aftermath of the 1999 fire disasters, the same civil servant had been assigned with the elaboration of new safety standards, this time going into the opposite direction. Instead of putting together a working group of experts and practitioners, he hired his friend, a brilliant scientist with good ideas, but absolutely no experience in tunnel ventilation design, for the elaboration of a new ventilation design guideline. The other members of the working group didn't know much about ventilation issues. Nevertheless, the elaborated draft was well worked out, with some useful requirements that would significantly increase the safety level, in particular the importance of controlling the longitudinal airflow in case of fire. Unfortunately, it was based merely on theoretical considerations without taking into account any thoughts about costs and benefits. The guideline was put in place without announcement and without any feedback from responsible designers and experienced practitioners. Only a few details were clarified by unofficial communication off the record. Since then, many issues have become obsolete due to new findings in the process of the application of the guideline and the ongoing technical progress. The responsible public servant, as well as his successor, have retired many years ago. But until today, the designers have had to work according to that standard. The responsible officials are obviously reluctant to learn. A new generation of engineers, most of them without previous experience, takes the requirements for granted. Many have never learned to think independently and work out technical solutions by applying logic, knowledge and knowhow. But obviously, this seems to meet a demand being made at the political level.

---

[35] Not the same one as mentioned in chapter 10.1, but they were working together.

Many years later, I was asked to work out a tunnel ventilation guideline for the road administration in another 'not to be named' country, where the European Union was about to invest huge sums in the development of traffic infrastructure. For peer review, a ventilation expert from a big engineering company was chosen. This company had strong relations to the local road administration, and was involved in most road infrastructure projects. Like most 'experts', the reviewer had extensive experience in calculations and simulations, but little practical understanding and absolutely no 'hands-on' experience.

My approach to work out a guideline was based on the principles as described in this book, taking into account the shortcomings of existing guidelines. I tried to heed the advice from George S. Patton (U.S. Army General in WW II, 1885 – 1945):

*Never tell people how to do things. Tell them what to do, and they will surprise you with their ingenuity.*

The requirements should be simple and understandable, described in a structured framework. Goals must be clearly stated. The part that is covered by many descriptive guidelines, the detailed technical concepts and design, should be left to the designers and performing contractors. They have the appropriate know-how. Most importantly, the goals were to be verified by a system of quality assurance, performance tests and acceptance procedures. This was applied both to design specifications and to operational standards. In a separate part a uniform quality standard for the equipment was defined by detailed technical specifications. In practice, small technical details make the difference between fulfillment and failure.

The reviewer agreed that the existing guidelines were focusing too much on design requirements, and should be simplified. Unfortunately, he did not really understand the importance of testing procedures and implementation details. Therefore, he worked out his own version of a guideline. Instead of contributing with some of his own conceptual ideas, he omitted the practical part, but adapted some questionable requirements from other existing guidelines. The client's representative finally accepted the reviewer's version – notably a document that had not been reviewed by any other expert, since I had withdrawn from the team.

A few years after the implementation of the new guideline, many tunnel projects have been planned and realized in that country, based on design and build contracts. Most local designers have little experience, and it becomes obvious that the guideline is full of flaws, as I had stated before.

Without going into the details, most equipment is partly oversized, but still does not fulfill its purpose in practice. A lot of time is wasted for technical discussions due to ambiguities and the lack of technical and functional requirements. Who is accountable for the disastrous situation?

By the way, the mentioned reviewer had also been the designer of the ventilation system of a long Alpine rail tunnel. That consisted of escape tunnels, which were equipped with an overpressure ventilation system to prevent smoke entering the emergency exits in case of fire in the rail tunnel. When the system was tested out by means of smoke tests, it showed that the overpressure ventilation even sucked the smoke from the rail tunnel into the escape tunnel. Nevertheless, the responsible designer blamed the unfavorable boundary conditions for this mishap. When I spoke to him years later, he was still convinced that a good engineer should rely on his design and that testing was not so important.

## 11.8  Sierre Tunnel Bus Accident

A tragic accident happened in the Sierre highway tunnel in Switzerland on March 13, 2012, when a Belgian bus crashed into the wall of a lay-by, killing 28 people, of which 22 were children. Another 24 children were injured. They were returning from their ski holidays in the Alps.

Fig. 40    Crashed bus in the Sierre highway tunnel   *(Public domain)*

The reason for the accident is unknown and will most probably never be discovered. Media attention decreased much faster than in case of the fire incidents, even when the number of casualties in this particular accident was higher than in almost all road tunnel fires. Unlike after the 1999 tunnel fires, which led to investments of several billions of EUR, the Swiss road authorities firstly even refused that there was a need for additional safety measures after the spectacular bus accident.

Lay-bys in tunnels are understood as a means of risk mitigation, enabling breakdown vehicles to leave the traffic lane inside the tunnel, thus avoiding subsequent rear-end collisions. It is difficult to estimate how many accidents have really been avoided by such lay-bys, which is an immanent characteristic of any risk mitigation measure. On the other hand, there is obvious evidence that lay-bys lead to a specific risk. Crashes into walls of lay-bys in tunnels happen regularly and are often fatal. Tragically, they have been also recognized as a popular way to commit suicide. While driving on a high-speed highway with 100 km/h, there are few other possibilities to crash into a perpendicular wall.

According to the Swiss tunneling standard, lay-bys would be foreseen only in tunnels with bidirectional traffic, where the permitted speed is lower and head-on collisions would be possible, but not in highway tunnels with unidirectional traffic, which is where that accident had occurred. In this respect, the representative of the Swiss road authorities has openly lied to the public by claiming that the Sierre tunnel would comply with the relevant standards.

In other countries, for instance in Austria and Germany, walls in highway tunnel lay-bys must be beveled or equipped with guidance elements to prevent direct impact. Those measures have proven their efficiency in many accidents. The Sierre tunnel was later refurbished with such guidance elements too.

## 11.9 Another Smoke Test

Once, in a 'not to be named' country, smoke tests in a highway tunnel with two tubes were performed. It is important to understand that in case of a fire in one tube, the traffic in both tubes must be stopped immediately. The moving traffic could draw the smoke into the non-incident tube, which serves as an escape route. But the project manager from the road authority did not want to close the tunnel completely and preferred to let a tube open to traffic while the smoke test would be conducted in the other tube. The designer wrote a warning letter that this would impose a serious risk, but was not heard.

According to design standards, the tunnel was equipped with a dividing wall on the tunnel portal, which was considered to be sufficient to prevent smoke spreading between the tubes.

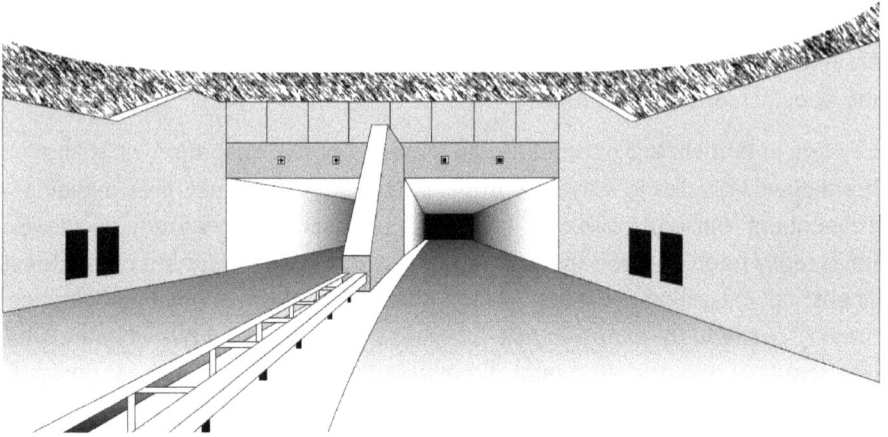

Fig. 41    Theory: Dividing wall according to design standard [7]

But during the test, smoke spread over the wall and seriously impaired the moving traffic. Luckily, no accident happened, and subsequently the traffic had been warned.

Fig. 42    Practice: Smoke spread during test

This has impressively demonstrated the limited usefulness of a measure that is still required by many design guidelines. A decade later, the issue had been examined in a research project, which basically confirmed what practitioners had known before.

By the way, the spread of smoke between tunnel tubes can be successfully avoided by appropriate operation of the tunnel ventilation system [109].

Fig. 43    The solution: Preventing smoke spread by appropriate ventilation

## 11.10 Down Under

Australia had always been a dream destination for me, based on an image of a wild country inhabited by Crocodile-Dundee type adventurers. When my personal circumstances allowed for it, I took the opportunity to work on some metro and road tunnel projects in Down Under, discovering a different engineering culture.

Vehicle emission standards in Australia have been far worse than those in other countries in the developed world and air pollution is a serious issue in the big cities. Subsequently, the exhaust air from Australian urban road tunnels must not exit the tunnel portals in order to protect the environment in the immediate vicinity of the tunnels. Instead, the contaminated air is extracted and blown by a vent shaft vertically into the atmosphere, where it is diluted to harmless concentrations. That way it had been practiced in many European urban road tunnels decades ago. Unfortunately, the extraction of air by powerful exhaust fans leads to a high energy consumption. Electricity in Australia is produced mainly in coal-fired power stations, causing the country to be the highest emitter of greenhouse gases per capita in the developed world.

Therefore, the goal of limiting some emissions in the vicinity of the tunnel portals led to a total increase of emissions in general. Obviously the decision makers didn't take into account such systemic considerations.

An alternative to blowing the tunnel exhaust air into the atmosphere would be the use of filters. However, what looks like an environment friendly solution at first glance rarely withstands a serious cost-benefit analysis. Given the fact that tunnels represent only a tiny part of the road network, the impact of filtration of tunnel air on the reduction of emissions in general is practically zero. Nevertheless, filters may provide a means to improve air quality in the immediate vicinity of road tunnels in specific situations. I worked out some measurements in such a test filter plant in an urban motorway in Sydney. Access to the site was quite restricted by formalized procedures, for instance I had to participate in a briefing on safety instructions. The safety officer demonstrated the use of an oxygen self-rescuer, which is commonly used in underground constructions, enabling you to breathe for a short time while evacuating from a smoke filled environment in case of a fire incident. On my question, the officer frankly admitted that he had never applied or even tested the emergency breathing apparatus in practice, nor had any of the workers. The devices are designed for single use and are quite expensive, and therefore never really tested, nor have the users ever been trained in practical use.

Before entering the air duct for the measurements, we also had to carry a Carbon Monoxide detector. In case of exceedance of allowable concentrations for health protection, it would sound an alarm and the contaminated environment had to be left immediately. Due to high emissions of noxious substances, particularly from the many trucks, the air in the duct was really bad, and the CO detector started to beep immediately. To be able to carry out the measurements, we simply had to switch the detectors off. I had to assess the risk of suffocation by my experience rather than relying on instruments and took the responsibility. Luckily, we survived without health impairment.

By the way, some Australian urban road tunnel projects were privately financed, based on a traffic prediction model. All of them went bankrupt, because traffic numbers fell short of the expectations. The drivers were reluctant to pay tolls for using the tunnel even if they could save a substantial amount of time. Before going bankrupt, some bankers and managers of construction companies had made high profit. Without the optimistic traffic predictions, the projects wouldn't have been realized at all, due to the lack of funds. Obviously, the investors had been deceived more than once. The same mistake of overestimation to fund project financing was repeated again and again.

Was it due to a deliberate failure to learn or simply because of prevailing commercial interests of particular stakeholders? When working in a design team for a new tunnel, I questioned the predicted traffic numbers, which were an important parameter for the design of the ventilation. Not only did those numbers exceed the lane capacity in the tunnel, but they were estimated up to 30 years ahead. Who can predict what kind of vehicles there will be in 30 years and even less in what numbers?

## 11.11 Tunnels and Warriors

A funny example for expensive, but malfunctioning equipment was installed in the Sonnenberg tunnel in Switzerland, which was built at the height of the Cold War in the 1970$^s$. Beside its main purpose as a bypass road below the city of Lucerne, the tunnel was designed to serve as a shelter for 20,000 people in case of a nuclear attack. For that reason, the tunnel portals were equipped with four heavy doors, each one weighing 350 tons to seal the inside of the tunnel hermetically. Only eleven years after the opening of the tunnel, an exercise was performed, in which the highway was closed and the tunnel was converted temporarily into a shelter. That test showed serious organizational deficiencies, and the doors could not be closed fast enough. Because of that, the whole shelter concept was in fact quite useless. Luckily, the cold war ended long ago, and Lucerne has never been subject to a nuclear attack. Meanwhile the doors have been dismantled during a tunnel refurbishment program.

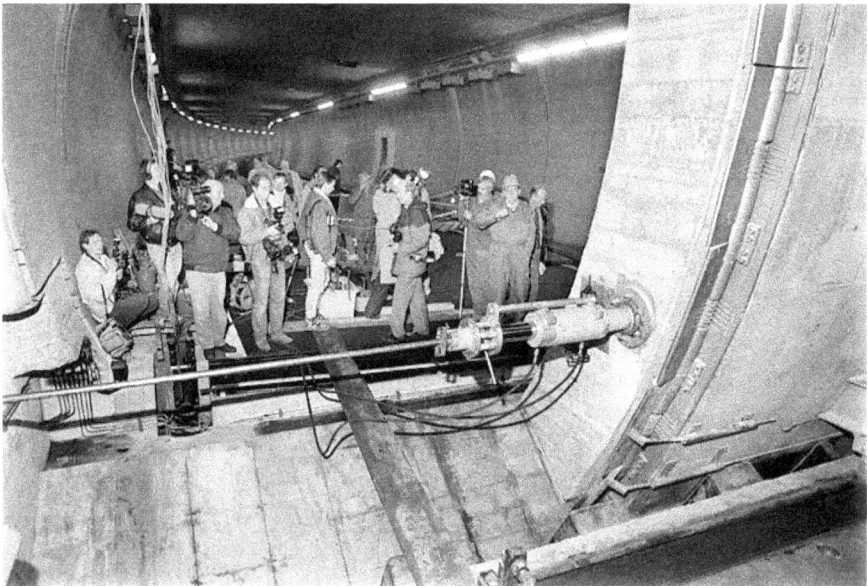

Fig. 44    Closing the Sonnenberg tunnel door during an exercise  *(Public domain)*

Many traffic tunnels in the whole world, both road and metro tunnels, have been designed for a second purpose, namely as emergency shelters or military bunkers, and some are still used in this role. Such provisions might be seen as a waste of funds in peaceful countries, but they provide preparedness for possible future wars or emergencies – if they work.

Tunnels seemed to be a favorite playground of 'cold warriors' in other aspects too. For instance, most bridges and tunnels in Switzerland and elsewhere were equipped with explosive charges. In case of war, deliberate destruction of infrastructure was strived for. When a traffic link could not be defended, the enemy should not be able to use it. During the fire incident in the Gotthard Road tunnel in 2001 those explosives imposed a serious additional hazard. By now, they have been removed from all Swiss tunnels. When I worked on a tunnel project in a central European country from the ex-communist bloc, the client required that the false ceiling, which provided the exhaust air duct, should be equipped with explosive charges so it could be demolished on demand. The only reason for that was to enable mobile rocket launchers to pass through the tunnel – more than a decade after the end of the cold war. Later the requirement for deliberate demolition was abandoned.

The Salang tunnel in Afghanistan, as mentioned in chapter 3.6, has been demolished various times by different warring parties in efforts to close the important link between Kabul and northern Afghanistan. However, even when the structure was heavily damaged, the tunnel itself stayed intact, and is subject to ongoing refurbishment programs.

Fig. 45    Outside the partly destroyed Salang tunnel in 2003    *(Reuters)*

# 12 Engineering Principles

In the previous chapters, many problems of risk management in practice were addressed. Here are some proposals to do it better, based on practical engineering principles. Such may be a matter of course for many professionals, but I have witnessed too many examples in which executives and decision makers inexcusably neglected these principles.

Since the rise of the industrial culture, engineers have been responsible for the technical progress. Bad engineers make mistakes, and good engineers make mistakes too. The latter try to take those mistakes into consideration, learn from them, and try not to make the same mistake twice. However, what was right in one situation might be wrong in another one. Faults happen inevitably and possible failures must be taken into account by conceptual provisions. Murphy, being an experienced aerospace engineer, showed that in his famous statement. Today it seems that many engineers waste most of their time and mental capacity with organizational and bureaucratic empty runs, rather than for analyzing, reasoning and 'engineering' within its original meaning.

For any development, design and implementation of technical systems, some general rules apply. Such principles might be useful for risk management in general:

1. Define the Goal
2. Gather Information and Analyze
3. Planning
4. Simplicity
5. Reserves
6. Backup, Redundancy and Safe States
7. Review, Testing and Improving
8. Better do nothing (than do the wrong thing)
9. Progress in small steps

## 12.1  Define the Goal

'*If you don't know where you're going, you'll wind up somewhere else*''
(one of the genius statements by U.S. baseball player and coach
Lawrence P. 'Yogi' Berra, 1925 – 2015)

The specification of goals is the first and most important step towards any endeavor. A goal must be understandable and verifiable, carefully defined as specifically as possible, leading to provable requirements. Deciding to achieve a goal means also to forgo other goals, which would be contradictory to the primary goal. Decision making might be a dynamic process, taking into account an ongoing progress and new information, and eventually omitting obsolete goals in favor of new ones.

A simple example for that is mountaineering: You cannot climb two mountains simultaneously. You can climb them one after another, but that requires time and additional resources. Or you might have to abandon your goal due to changing circumstances, for instance adverse weather. Then alternative goals must be considered. Defining a goal is only useful when it can be realistically achieved, and achievement can be proven with reasonable efforts. Goals that are not verifiable by testing and measuring are practically useless.

The definition of goals is not a matter of course, as Albert Einstein (Physicist and Nobel prize winner, 1879 – 1955) had stated:

'*Perfection of means and confusion of ends seem to characterize our age*'

Many technical standards prescribe technical solutions and design criteria without stating clear safety goals to be achieved. This leads to regular confusion, particularly when the requirements prove to be either contradictory or not reasonably feasible. In such cases, the definition of goals has been wantonly neglected.

## 12.2  Analysis and Information

The analysis of a problem, based on useful information, would be essential before taking any further steps. Unfortunately, to analyze and understand the whole system can hardly ever be achieved. The necessary information is incomplete, and there are unknown aspects. Perception and processing of this information is subject to biases. Many contradictory arguments have to be taken into account. Even the definition of the system in a particular context is a difficult task. You must limit the system unless you get lost in infinity, but by fixing limits, you exclude factors that possibly might have an influence.

Even within the limits, you will never be able to perceive and determine all relevant aspects, since many are unknown. A proposed approach for system analysis is to ask the following questions, taking into account that the answers may vary with time and circumstances:

- What is the problem?
- What are possible solutions?
- What are the benefits and disadvantages of each solution?
- What are the costs?
- What can happen? When and how can it fail?
- What are the possible consequences of recognized failures?
- What are the possible probabilities of recognized failures?
- What can be done to prevent recognized failures?
- What are the costs of prevention measures?
- What can be done to mitigate damages of recognized failures?
- What are the costs of damage mitigating measures?
- How robust is the system against the unforeseeable?

and, taking into account the social context of different stakeholders:

- Who decides?
- Who profits?
- Who suffers from failures?
- Who bears responsibility for failures?
- Who pays for the investment?
- Who pays for the operation?
- Who pays for damages?

## 12.3  Planning

Carl von Clausewitz wrote:
*'Don't do the first step before having planned the last one'*

Before understanding, you should not plan, and before planning, you should not act, at least when significant risks may be involved. That's a good advice in theory, but only half the story. The relevant issues have been described in chapter 7.1.

## 12.4 Simplicity

*'Truth is ever to be found in simplicity, and not in the multiplicity and confusion of things'* (Isaac Newton)

The principle of choosing the simplest variant has been known since medieval times as Ockham's razor. Among engineers, it is known as KISS (Keep it simple ...[36]). Risks of failure increase with the complexity of a system unless the system is implemented and tested with adequate effort. There is a limit.

The design, construction, implementation, testing, operation and maintenance of systems are economically not feasible when they exceed a certain level of complexity. Simplicity, reliability and testability are closely connected with one another. With increasing complexity of technical systems, the error rate increases dramatically, and the efforts to test and improve rise disproportionately in order to achieve adequate reliability and freedom from error. Complex systems can reach a high level of safety, but only at enormous costs. That's why for instance spaceships and nuclear power plants are so expensive.

Depending on the system, there may be a point of maximum safety, beyond which safety cannot be increased further even with enormous additional effort.

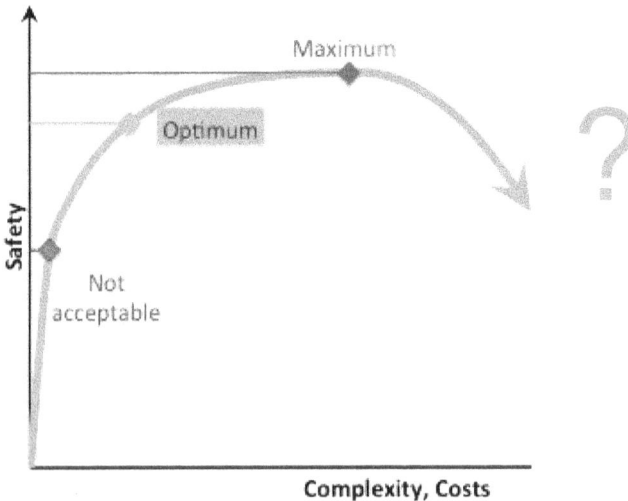

Fig. 46    Safety against complexity

---

[36] The meaning of the word with the second 'S' is left to the reader.

Simple solutions are not always useful. Too complex solutions can't even be understood and assessed regarding their usefulness. Therefore, concepts and systems should be kept as simple as possible as long as they fulfill the goals and requirements. In contrast, if complex things have been built in large numbers and work for a long time, their reliability increases as a consequence of an evolutionary process of trial and error. Mass product electronic devices, computers and automobiles have reached an amazing dependability despite their complexity.

Whether a technical system works and complies with the requirements is not a matter of course. It takes a lot of effort, and that increases with complexity. Laymen, academics, theoreticians and managers without hands-on experience in designing and implementing a system in practice regularly underestimate the necessary effort and resulting risks to make a technical system work. It is obviously much easier to get lost in confusing, complex specifications than to work out simple, useful concepts. Simplification is hard strenuous work and must be paid. Many representatives of awarding authorities don't understand that and prefer to reimburse engineering in relation to project costs and the amount of produced paper. There are many official engineering fee bases according to project complexity and construction costs. Whether such bases are a useful motivation for good engineering practice might be questioned.

The importance of simplicity applies not only to technical systems, but also to standards and rules. The legal system in many societies has been blown out of proportion. The legal industry profits from regulations that are beyond comprehension by most citizens. Unfortunately, this may lead to ignoring the laws.

However, the KISS principle must not be misunderstood. There is a limit to simplicity. A too simple system which might not achieve the goal is useless. The achievement of goals is of utmost priority. Therefore, some systems may become extraordinarily complex all the same. In some cases the usefulness of defined goals might be questioned as well.

## 12.5 Reserves

*'If you do not expect the unexpected you will not find it, for it is not to be reached by search or trail'*
(Heraclitus, Greek philosopher, 535 – 475 BCE)

Unlike accountants, engineers work with uncertainties. Any design process is based on models, estimates and assumptions. Theorists tend to believe too much in the models. Reality is always different from what you have presumed previously and to take this discrepancy into account, reserves must be provided to take into account the unforeseen.

There is no such thing as exact design. Any technical system and equipment must be oversized to take into account uncertainties about calculation models, boundary conditions and overlooked factors. Achieving exactly the target value means that you were lucky and on the brink of non-compliance of requirements. Planning an undersized system or equipment means not accomplishing the task, therefore probably not getting paid, losing your reputation, and even being sued for paying compensation. The question is not whether but to which extent to oversize. This applies to technical safety margins as well as to financial and time reserves. The name 'safety margin' implies that its purpose is to increase safety. Reserves provide flexibility for changing conditions and unexpected events. Unfortunately, reserves bind precious time and 'idle capital' and must be limited. Reserves in one machinery may require extensive additional measures on the whole system. For instance, oversizing a ventilation fan by 25% in volume flow doubles the necessary power. That leads to more expensive mains cables, transformers, switchgear, and consequently even bigger power plants, if not taking the energy at the costs of other power consumers. Alternatively, you may enlarge the ventilation ducts, which then would increase construction costs.

More of the good is not always better. Excessive reserves may even bring additional risks and increase complexity. For instance, smoke extraction in highway tunnels, which is seen as an additional safety measure, does in fact increase the risk to the tunnel users, as was explained in chapter 11.6. Lewis mentions the example of airplanes, in which increased safety margins for structural strength would obstruct the ability to fly [80]. Conservative assumptions do not always lead to a conservative outcome. In practice, reserves are determined by rules of thumb, mostly based on experience, and therefore they are prone to the problem of induction. There is no optimal determination of reserves, merely a weighing up of costs against risks. Even with excessive oversizing, an unexpected event may cause failure.

## 12.6 Backup, Redundancy, Safe States

Murphy's Law is not about the unexpected. On the contrary, it states what has to be expected: Humans do err and technologies do fail. Technical systems must be tolerant about faults. Appropriate fallback plans and redundancy requirements must be defined taking into account possible failure scenarios. When plan A fails, then plan B comes into operation. For important systems, a plan C must be able to absorb the failure of plan B, possibly a plan D will be required and so on. Finally, if all plans fail, then the system should at least reach a safe state without serious damage, preventing an uncontrollable disaster. Impossibility to comply with this requirement is a basic drawback for technologies with the potential to cause a disaster, for instance nuclear power generation.

On the other side, redundancy significantly increases the complexity and hence also the cost, time and effort for appropriate measures to ensure reliable functionality of the system. Simple systems with redundancy requirements get complex. Complex systems with redundancy requirements may get uncontrollable, as was described by Perrow [97], which is confirmed by the experience of professionals who have worked on large projects. In practice, redundancy requirements can in fact lead to an increase in failure probability. This is one of the many contradictions in life, therefore an acceptable compromise has to be found. Most of us have experienced the consequences of loss of data at work and in private life. In a professional environment, providing data backup is a matter of course. However, backup requirements may increase the complexity and costs of the whole IT infrastructure, and slow down the working process. In fact, IT security may become a massive impediment to efficient work.

Critical technical systems are equipped with a power supply backup in form of an uninterrupted power supply and a diesel generator to ensure continuous operation or at least a controlled shutdown, which is essential for many applications. Unfortunately, as a matter of principle, even the most sophisticated backup systems may fail. Diesel generators have to be tested out regularly, which subjects them to wear and limited lifetime, and they constitute another technical component that may fail or cause accidents.

Automated systems can fail or be overridden by incorrect manual operation, deliberately or accidentally. For any failure scenario, safe states are to be defined, which must be reached even in case of a complete system breakdown, for instance in case of a loss of power supply.

Similarly, the design of constructional and operational provisions should ensure that even in critical states, in which the foreseen function is not guaranteed, at least failures with catastrophic consequences can be avoided. For instance, the forces defining the strength of constructional measures exceed the pressures that have to be taken into account for the design of a ventilation system. Buildings must not collapse, even when the failure of equipment is accepted in extreme cases.

Once I worked on the development and application of industrial waste air treatment plants. One of the clients was a big international production company. The contaminated exhaust air from the production process was molesting the residents and was not good for public relations. The clients project manager brought the most important requirement to the point: The performance criteria of the waste air treatment plant must be achieved sooner or later and possible flaws must be corrected to fulfill the contract. But that was only of second importance to the client. What he really worried about was that the air treatment in the production line must not under any circumstance affect the production process. Downtimes would cost millions of dollars per day, and assuming liability would have led to the bankruptcy of the equipment provider, which was my employer. Therefore, a safe state had to be implemented and thoroughly tested out, before the commissioning process of the waste air treatment could proceed. The safe state was in fact given a higher importance than the system performance. Such an approach is proven practice in many fields of system engineering.

For instance, according to the EU Directive [34] all equipment with potential hazard must be equipped with an emergency button to bring it into a safe state, usually by switching it off. This button must be easily visible and reachable, even under stress conditions. Where a safe state is not achievable, an unpredictable risk is intrinsic.

## 12.7   Review, Testing and Improving

*'Though experience be our only guide in reasoning concerning matters of fact'* (David Hume, Scottish philosopher, 1711 – 1776)

'Trial and error' has been the mechanism of technical progress and basically of the whole evolution of life. Analysis, design and calculations, the tools of engineering, are a shortcut, but not a replacement for this process. Would you apply a medicament that has not been previously thoroughly tested on animals and volunteers? Would you fly with an airplane of a new type that proved itself in computer simulations but has never been in the air before?

Does a machinery really work reliably or has it never been tested under stress conditions? The same applies to personal abilities. Have you proven your skills under difficult conditions in the real world, or have you shown your abilities only in training and simulation in a protected environment?

According to Karl Popper (Austrian / British philosopher, 1902 – 1994), no theory in science can ever be proven, only falsified [106]. Progress of scientific knowledge is achieved by attempts to falsify the postulates and theories. A theory that withstands falsification might be designated 'right'. However, there is no guarantee that it might not be falsified one day. The principal intention of experiments is to falsify a postulate, not to prove it. When something cannot be falsified in principle, or it is not a consequence from a falsifiable statement, then it has no implication in reality, being in the realm of pseudo-sciences.

Without having read Popper, many engineering practitioners use this principle when developing a new system, technology or machine to improve its safety and reliability. Errors can be found and eliminated only by thorough review and testing under stress conditions. The right question is not: 'Does it work?' but 'When and how does it fail?' The purpose of testing is not to demonstrate that everything works fine, but to find faults that are to be eliminated. If no faults are found, the testing might not have been thorough enough. Requirements must be verifiable. What is not testable bears an inherent risk. Not only the design conditions have to be tested, but additional stress should be applied to test critical states – as long as possible failures don't lead to unacceptable damage. When a system is tested out for really critical states that will hardly ever occur in operation, then it can be assumed that it will work under all probable circumstances.

The earlier faults are found, the easier and less expensive corrections are. In an early stage, the peer review of design documentation can be seen as a kind of test as well. Validation of calculations and simulations is essential. Not only to err is human, but also to be blind to one's own mistakes and to focus on the mistakes of the others. That's why the dual control principle is required by any serious quality assurance system.

For testing and subsequent improving, appropriate time and funds are required. Many design rules and basic clients' requirements that have the greatest influence have never been seriously reviewed. Later, in the ongoing design process, designers and executing contractors have to adopt inadequate and botched concepts. Then the project documentation is peer reviewed, but the big mistakes have been made before.

However, review is only about paper, but what counts in practice is whether it works in reality or not. Thorough testing is a matter of course in applications where failures would be evident and lead to substantial financial damage, for instance in power plants and production facilities. Reliable functionality is essential and given highest priority. On the other hand, testing is often neglected in equipment that is not required for the primary production process and therefore does not generate revenue. Safety systems, like for instance fire protection equipment, are never or very rarely used, and actually nobody knows whether they really work, unless they are regularly tested. Since unexpected conditions cannot be defined by principle, engineers may use all their imagination to evaluate realistic and critical scenarios. Unfavorable boundary conditions should be simulated as realistically as possible.

Fig. 47     Simulation of adverse boundary conditions by a mobile fan in a tunnel test

Technical systems that are produced in large series, for instance cars or computers, achieve a high level of reliability despite their complexity. The sheer number of devices that are used in daily application provides a significant amount of testing scenarios. Even then, failures happen. In contrast, in system engineering, each system, for instance a production plant, might be considered as a prototype.

Just as testing of technical systems increases their trustworthiness and reduces risks, hands on experience improves abilities and confidence of professionals, particularly those that have to perform in high risk situations under stress, for instance emergency services. Experience is acquired by trying it out and regular training. If feasible, the stress level in training should be worse than can be expected in real operations. Making errors must hurt, but not be disastrous. In fact, you learn more from errors than from success. But for that, you need the ability to admit errors. Many people ignore especially their own mistakes, but to err is human and errors always occur. The testing of safety equipment is often omitted, not only under financial and time pressure, but out of sheer short-sighted convenience.

Testing is not reality, but should simulate it as feasibly as possible, without inflicting the damage of a real incident. When a test could lead to unacceptable damage, it is not feasible. That aspect is essential, and the very challenge of test engineering. Unfortunately, testing and maintenance is an intervention in a functioning system and may be an additional source of errors. Many disasters were caused by maintenance errors or inappropriate testing procedures. For instance, Chernobyl was in fact the consequence of a test scenario that was not really thought out in that respect.

## 12.8 Do Nothing

'無爲 *Wu Wei*' (in English: '*If it ain't broke, don't fix it*')

Sometimes it would be wiser to simply avoid doing the wrong thing, instead of doing anything. Doing nothing does not mean not making a decision. It means quite the opposite, it is the decision not to interfere and let other forces do the job. In this respect, doing nothing may be a powerful way of acting. Humans tend to useless eagerness, blind activism and quick-fix measures. Immediate threats often lead to a pressure for action, without consideration of what would be appropriate. If decision-making about such events has not been planned before, particularly in case of unique incidents, the response might possibly not be adequate.

For instance, when a fire in a tunnel occurs, it would often be wiser to switch the forced ventilation off and let the smoke spread by natural forces. Unfortunately, operators tend to switch the ventilation in one or the other direction, especially on demand from emergency services that want to get access to the fire site. By doing this, they expose tunnel users on the other side to the smoke, possibly even killing them, as is described in chapter 11.2.

When there is no urgent hazard, there is enough time for reasoning, thorough analysis and decision making. As we usually focus on the issues with high consequences, we neglect the other ones, automatically doing nothing about them. Low risks are not worth it, since many problems solve themselves as time passes by.

## 12.9   Progress in Small Steps

The goal of 'trial and error' is finally success. Each step ahead bears a chance, but also an unknown risk. Failures will occur, and a step back has to be taken occasionally.

You can only work efficiently when repeating the same thing again and again, learning from previous experience. When doing something new, unforeseeable effects may occur. Practitioners with hands-on experience know that intuitively, and work very cautiously in small steps, testing and analyzing with checks and counter checks after each step. Some errors would lead to extensive damage, and some cannot be corrected at all. Taking sufficient time is essential, but often contradictory to prevailing economic conditions. Apart from that, slow progress might be psychologically frustrating, since going ahead is a natural human impulse.

Sometimes a sensational invention may seem to be a big step ahead, but the implementation may reveal formerly unknown risks. Big step inventions require small step implementation, with diligent testing at each step. Chances may be revealed too: Previously unknown secondary effects of inventions may become more important than the original purpose.

Revolutions, implied as a deliberately planned process of big steps are risky and often worsen the situation, whereas evolutions of small steps have a better chance of leading to sustainable development. In this context, the term 'industrial revolution' is misleading, since it has been rather a developing evolutionary process with two steps ahead and one step back, which is still going on.

# 13 Summary

The next disaster will happen inevitably. Probably it won't be foreseen, but rather it will be an incident that has been expected or has even happened before. After the incident, the situation is mostly quite safe, but that's when people start to think about precautions. High media coverage will lead to a hype and this will increase profit for the media themselves. Politicians will seek to shine by demanding new safety measures without taking into account any cost benefit considerations. But what has really happened? Was the safety concept – if there was any – based on a thorough analysis and experience, or rather subject to random decisions of dubious 'experts'? Were basic safety rules correctly applied? Has the existing safety equipment been maintained and regularly tested? Was the disaster caused by an operator who had to work under constant pressure without proficient training, being sleep-deprived, exhausted and distracted by information overload? Could the extent of damage have been reduced by appropriate behavior of the affected?

But what are the risks that we really face? Most likely you will die at an old age by heart disease or cancer[37]. A few generations ago, most people died because of contagious diseases. Many children did not survive the first years of life. Today, those once feared diseases have been extinguished or massively pushed back by better nutrition, vaccine, hygiene and medical treatment. But improved health care may have unwanted side effects, like the massive overpopulation in many developing countries.

Take into account the alternatives to the risks that you fear. For instance, I am regularly involved in discussions about risks in tunnels. Do you fear driving through a tunnel in the mountains? Without the tunnel, your car might be possibly swept away by an avalanche. Tunnels in cities make life for the local residents much safer and comfortable. Unfortunately, some tunnel projects are delayed or even cancelled because new safety standards lead to massive cost increase. Simple analysis would show that many of those requirements only marginally influence the safety of tunnel users, and some may even increase risks, as is explained in chapter 11.

---

[37] This statement is based on estimated probabilities for the whole population according to statistical data. Of course, your particular situation may be completely different.

Providing another example, many citizens are concerned about pesticides and genetically designed food. But in the past, hunger and starvation were regular threats. People were poisoned by spoiled food, poisonous plants or rotten meat. Complete losses of harvests occurred occasionally, like for instance the one in Ireland in the middle of the 19$^{th}$ century due to potato blight, when more than one million people died and another million had to emigrate. Even with a guaranteed supply of potatoes or wheat, a one-sided diet based on starch imposes a significant health risk.

Are you afraid of terrorism? Violence from your fellow citizens has been one of the most significant threats in the whole human history. In the 20$^{th}$ century, dozens of millions of innocent people were killed by orders from their own government in ruthless dictatorships. Young men were sacrificed in futile wars. In traditional hunter-gatherer societies, the risk of a violent death was even higher, due to regular skirmishes, mutual raids and murders. Today, despite of occasional outbreaks of violence by misguided maniacs and the following scaremongering in the media, we should be grateful to live in a society that is as peaceful as never before.

Global warming, partially caused by the combustion of fossil fuels, may impose a serious threat to many societies in the long run. But consider that before the discovery of fossil fuels, wood was used as primary fuel. Huge landscapes were deforested, and consequently fertile soil was forever lost to erosion. Deforestation led to avalanches in the mountains and floods in the lowlands. In fact, fossil fuels contributed fundamentally to the industrial revolution and to our present wealth and safety.

Nuclear contamination is feared as a major risk in public opinion. But natural radiation, for instance by radon emissions from soil or solar radiation, kills more people every year than any nuclear disaster. Nuclear waste remains a threat for thousands of years, but imposes a negligible risk, especially in comparison with the ecological disasters that our ancestors have caused. Remember the extinction of many species of animals, the loss of biodiversity, or the deforestation and desertification in many parts of the earth that we have to get along with.

The list could be continued endlessly, and some of the mentioned issues are examined in more detail in this book. Human risk perception and decision making are mostly based on gut feelings, bias, superstition and media attention. Analyzing and thinking critically impose hard work, and unfortunately are no guarantee for success anyway.

Our ancestors had little influence on the risks that they faced but did not understand. Today, we are able to increase our safety and wellbeing, based on scientific, medical and technological progress, and risk management. Useful risk management is not about maximizing safety, but finding the right balance between risk and safety, development and consolidation, costs and benefits. In that process, we must be aware that many, even crucial factors are beyond our knowledge, out of our control and often merely coincidental. Absolute safety is never achievable, since two main factors imply unpredictability: Human behavior, which is often unforeseeable, and the element of simple chance and hazard, which is inherent in nature.

Many disasters don't result from basically unknown risks or faulty risk management, but from risk management that was not applied correctly and from obviously wrong decisions. But that's easy to state with the benefit of hindsight. Analytical methods on risk evaluation are sophisticated, but remain limited in principle and are often not accepted by the general public. Risks with infinitesimal probability and high potential impact cannot be determined. Even worse, the most severe risks might be unique events that nobody has thought about before.

Big steps ahead were taken in the prevention and mitigation of foreseeable risks. Technical and medical progress, together with almost unlimited availability of energy sources mainly by fossil fuels, have led to an unprecedented level of wealth, prosperity and safety. But there is still a demand for improvement. A transition to a more sustainable supply of energy and raw materials would be the most important task of the future. Only further progress will ensure wealth and safety in the long run. Any progress has intrinsic risks and should be done in small steps. The bigger the steps, the higher the risk. Errors will occur and incidents will happen, but the resulting damage must be kept on an acceptable level. The probability of failure and subsequent risks can be reduced or even minimized by appropriate measures, but for that, a price has to be paid. Safety is not available for free. Time and money have to be invested into safety measures and equipment, implementation and thorough testing, but also the price of missed opportunities by avoiding risks has to be taken into account.

From an engineering point of view, many problems could be technically solved, but psychological and societal aspects prevent us from doing so. Humans are social beings, which can survive only within a society, but the inherent patterns and mechanisms of social interactions can become a significant source of risk.

Many threats originate from our fellow human beings, deliberately or accidentally. Humans strive for the gratification of particular interests, being willing to sacrifice their safety in the long run and burden unknown risks to their fellow human beings and even their successors.

On a societal level, as far as the collective risk is concerned, we must bear in mind that safety is always a trade-off against other important priorities, affecting stakeholders with contradictory interests. Who pays, and who profits? Since there is a mismatch between objective risk, subjective risk perception and personal interests of stakeholders, decisions made by economic and political leaders lead rarely to good solutions in terms of view of useful allocation of limited funds. Safety is an investment without guaranteed revenue. Disasters that don't happen don't lead to immediate costs. However, when they happen, the consequences might exceed any expectations, and the resulting costs are not always reimbursed by the perpetrator or covered by insurance policies. The ultimate insurance is always the society. We all have to pay up to a certain extent. This has a very strong political implication. In a democracy, the definition of an acceptable risk level must not be left to public servants with almost unrestricted power of decision, often pursuing hidden economic interests. Bias, willful blindness, foolishness and mental corruption are common human characteristics, which must be taken into account in any risk issues.

Overcoming dogma and superstition, logical reasoning and empirical evidence lead to new findings and scientific progress. By that, we have reached an unprecedented level of wealth and safety, mostly accompanied by a political development towards a liberal society, in which citizens have got the freedom to choose their way of life within a broad range of boundary conditions. Open-mindedness and honest communication, taking into account different viewpoints, would be essential. Failure to admit errors and reluctance to learn and change your mind does significantly increase the risk in a dynamic environment that is to a large part unknown. Ideologies and dogmas that are not to be questioned were fine in the Dark Ages. People with too much confidence in their beliefs, not taking into account human fallibility, were often chosen as leaders in traditional societies. Certainty calms nerves – uncertainty makes you feel uncomfortable. However, circumstances are different in a modern, quickly developing society with new technologies, where unjustified certainty increases exposure to risks.

Today if somebody claims to be 100% sure about any issue, he is either a liar or an idiot. It is better to bear in mind Bertrand Russell's statement:

'*I would never die for my beliefs because I might be wrong*'

Or with the words of Socrates (Greek Philosopher, 469 – 399 BCE):

*'I know that I know nothing'*

Competent, motivated professionals who understand what they do, being aware of possible (and inevitable) mistakes in their work, taking responsibility for doing a good job, having sufficient time and a fair reimbursement, would be a guarantee for good workmanship, quality and a high level of safety. Useful risk management would also be about preventing permanent stress and depletion, bungling, ignorance, incompetence, dilution of responsibility and useless bureaucracy in organizations.

Rules and standard procedures are useful when quick and efficient action is essential, for instance in situations of imminent danger when damage has to be limited. In contrast, decisions must not be made under pressure when it is not necessary. 'Consulting your pillow' is a wise strategy. Being hurried, exhausted, sleep-deprived, distracted, intoxicated, or under social pressure leads to mistakes, resulting in possibly unacceptable risks. Pressure for quick action increases the probability of doing the wrong thing. Reality will always be different from what you have thought before.

Any change leads to new opportunities, chances and risks. Despite mandatory safety standards, implied by superiors, laws or insurance companies, you – as an adult person and responsible citizen – are finally responsible for your own safety. At least you should ask yourself where your desired level of safety is and what you are willing to pay for that. In financial investment, it's quite obvious and simple: The higher the interest rate, the higher the probability that you lose everything. Contradicting promises are a fraud. Real life is more complicated, but the basic principle is the same. Finally, risk management is mostly about personal attitude, focusing on the following topics:

- General skills, knowledge and know-how
- Awareness of uncertainty and limits of knowledge
- Ability to gather, assess and analyze necessary information
- Open mind, considering different opinions
- Admittance of mistakes
- Strive for continuous questioning, testing and improvement
- Awareness of sources of error, like time pressure, distraction, social pressure, and blind confidence
- Willingness to invest in reserves and redundancy
- Understanding Murphy's Law
- Acceptance of residual risks

This book shall end with a short reflection about a popular proverb.

*'It doesn't matter how many times you fall ... what matters is how many times you stand up'*

There are many old and new sayings with a similar meaning. Progress is mostly three steps forward and two steps back. Falling down as a symbol for failure is a key ingredient in the process of trial and error. Every fall can teach you a lesson. But there are other aspects to consider too. Falling down and standing up costs time and energy, physically and mentally. Falling too many times may be grueling and finally wears you down. One fall might be your last one. Therefore, try to think ahead of what the consequences of falling may be and choose your falls carefully. If you have no choice, I wish you good luck!

Fig. 48    Fall with acceptable consequences

# 14 Appendix

## 14.1 Bibliography

[1]     Ariely D., Predictably Irrational: The Hidden Forces That Shape Our Decisions, 2008

[2]     Arrow K., The Limits of Organisation, 1974

[3]     Asch S., Opinions and Social Pressure, 1955

[4]     Ashenfelter O. Greenstone M., Using Mandated Speed Limits to Measure the Value of a Statistical Life, 2002

[5]     Atkins D., Human Factors in Avalanche Accidents, 2000

[6]     Aven T. et al., Uncertainty in Risk Assessment: The Representation and Treatment of Uncertainties by Probabilistic and Non-Probabilistic Methods, 2014

[7]     BD 78/99 Design Manual for Roads and Bridges, Design of Road Tunnels, UK Highway Agency, 1999

[8]     BEA, Final Report on the accident on 1st June 2009 to the Airbus A330-203 registered F-GZCP operated by Air France flight AF 447 Rio de Janeiro - Paris, 2012

[9]     Bechmann G., Risiko und Gesellschaft: Grundlagen und Ergebnisse Interdisziplinärer Risikoforschung, 1997

[10]    Bedford T., Cooke R., Probabilistic Risk Analysis: Foundations and Methods, 2001

[11]    Berns G. et al., A shocking experiment: New evidence on probability weighting and common ration violations, 2007

[12]    Berns G. et al., Neurobiological Correlates of Social Conformity and Independence During Mental Rotation, 2005

[13]    Bernstein P., Against the Gods: The Remarkable Story of Risk, 1996

[14]    Bertrand M. et al., Does corruption produce unsafe drivers? 2006

[15]    Black W., A Comparative View of the Mortality of the Human Species at All Ages; and of the Diseases and Casualties by Which They Are Destroyed or Annoyed, 1788

[16]    Blalock G., Kadiyali V., Simon D. H. Driving Fatalities after 9/11: A Hidden Cost of Terrorism, 2009

[17]    Borodzicz E., Risk, Crisis and Security Management, 2005

[18]    Bostrom N., Existential Risks: Analyzing Human Extinction Scenarios and Related Hazards, 2002

[19]    BP Statistical Review of World Energy, 2015 (bp.com/statisticalreview)

[20] Broadribb M., Lessons from Texas City: A Case History, 2006

[21] Cashdan E., Risk and Uncertainty in Tribal and Peasant Economies (Westview Special Studies), 1991

[22] Chapman C., Durda D., Gold R., The Comet/Asteroid Impact Hazard: A Systems Approach, 2001

[23] Chapman C., The hazard of near-Earth asteroid impacts on earth, 2004

[24] Chiles J., Inviting Disaster: Lessons From the Edge of Technology, 2001

[25] Clark G., A Farewell to Alms: A Brief Economic History of the World, 2007

[26] Covello V., Mumpower J., Risk Analysis and Risk Management: An Historical Perspective, 1985

[27] Damasio A., Descartes' Error: Emotion, Reason and the Human Brain, 1994

[28] De Atkine N., Why Arabs Lose Wars, 1999

[29] Dekker S., The Field Guide to Understanding Human Error, 2006

[30] Diamond J., Collapse: How Societies Choose to Fail or Succeed, 2005

[31] Diamond J., The Third Chimpanzee: The Evolution and Future of the Human Animal, 1992

[32] Diamond J., The World until Yesterday: What Can We Learn from Traditional Societies? 2013

[33] Directive 2004/54/EC on minimum safety requirements for tunnels in the Trans-European Road Network

[34] Directive 2006/42/EC on machinery

[35] Dörner D., Die Logik des Misslingens: Strategisches Denken in komplexen Situationen, 1989 (The Logic of Failure: Why Things go wrong and what we can do to make them right, 1996)

[36] Douglas M., Wildavsky A., Risk and Culture, An Essay on the Selection of Technological and Environmental Dangers, 1982

[37] Duffé P., Marec M., Cialdini P., Rapport Commun des missions administratives d'enquête technique Française et Italienne relative a la catastrophe survenue le 24 mars 1999 dans le tunnel du Mont Blanc, 1999

[38] Dyatlov A., Why INSAG has still got it wrong, Nuclear Engineering International, 1995

[39] EPA Assessment of Risks from Radon in homes, 2003

[40] Fischhoff B., Risk Analysis and Human Behavior , 2012

[41] Fischhoff B., Slovic P., Lichtenstein S., How Safe Is Safe Enough? A Psychometric Study of Attitudes Towards Technological Risks and Benefits, 1978

[42] Fleiss J., Palan S., Of Coordinators and Dictators: A Public Goods Experiment, 2013

[43] Fogel R., The Escape from Hunger and Premature Death, 1700-2100: Europe, America, and the Third World, 2004

[44] Fryer B., Sleep Deficit: The Performance Killer. A Conversation with Harvard Medical School Professor Charles A. Czeisler, 2006

[45] Gaissmaier W. Gigerenzer G. 9/11 Act II: A Fine-grained Analysis of Regional Variations in Traffic Fatalities in the Aftermath of the Terrorist Attacks, 2012

[46] Gardner D., Risk: The Science and Politics of Fear, 2008

[47] Gigerenzer G., Calculated Risks: How to Know When Numbers Deceive You, 2002

[48] Gigerenzer G., Dread Risk, September 11, and Fatal Traffic Accidents, 2004

[49] Gigerenzer G., Gut Feelings: The Intelligence of the Unconscious, 2007

[50] Gonzales L., Deep Survival: Who Lives, Who Dies, and Why, 2003

[51] Heffernan, M., Willful Blindness: Why We Ignore the Obvious at Our Peril, 2011

[52] Hill K., the Psychology of Lost, 1998

[53] Holt A., Review of the Progress of Steam Shipping during the last Quarter of a Century, Minutes of Proceedings of the Institution of Civil Engineers, 1877

[54] Hopkins A., Failure to Learn: The BP Texas City Refinery Disaster, 2008

[55] INPO 11-005, Special Report on the Nuclear Accident at the Fukushima Daiichi Nuclear Power Station, 2011

[56] INSAG-17, Independence in regulatory decision making, IAEA, 2003

[57] INSAG-7, The Chernobyl Accident: Updating the INSAG-1, IAEA, 1992

[58] ISO 12100:2010, Safety of machinery: General principles for design, risk assessment and risk reduction

[59] ISO Guide 73:2009, Risk management - Vocabulary

[60] ISO Guide 78:2012, Safety of machinery - Rules for drafting and presentation of safety standards

[61] ISO/IEC 31000:2009, Risk Management - Guidelines for principles and implementation of risk management

[62]   ISO/IEC 31010:2009, Risk management - Risk assessment techniques

[63]   Jaeger C., Renn O., Rosa E., Webler T., Risk, Uncertainty and Rational Action, 2001

[64]   Janis I., Groupthink: Psychological Studies of Policy Decisions and Fiascoes, 1982

[65]   Jenssen G., Human Factor and Behavior: What we can learn from dealing with real road tunnel fires in Long Single-Bore Tunnels, 2015

[66]   Kahn M., The Death Toll From Natural Disasters: The Role of Income, Geography, and Institutions, 2003

[67]   Kahnemann D., Thinking, Fast and Slow, 2011

[68]   Kahnemann D., Tversky A., Prospect Theory: An Analysis of Decision under Risk, 1979

[69]   Kitzinger J., Researching risk and the media, 1999

[70]   Kletz T., An Engineer's View of Human Error, 2001

[71]   Kletz T., What Went Wrong? Case Histories of Process Plant Disasters and How They Could Have Been Avoided, 2009

[72]   Kohn L., Corrigan J.M., Donaldson M.S.(Editors), To Err is Human: Building a Safer Health System, Committee on Quality of Health Care in America, Institute of Medicine, 2000

[73]   Kopelman P., Health risks associated with overweight and obesity, 2007

[74]   Kruger J., Dunning D., Unskilled and unaware of it: how difficulties in recognizing one's own incompetence lead to inflated self-assessments, 1999

[75]   Kuklinski J., Misinformation and the Currency of Democratic Citizenship, 2000

[76]   Kuran T., Sunstein C., Availability Cascades and Risk Regulation, 1999

[77]   Lammers J., Stapel D., How power influences moral thinking, 2009

[78]   Larson S., Human Factors in Avalanche Incidents, 2006

[79]   Lewis H., How Safe is Safe Enough?: Technological Risks, Real and Perceived, 2014

[80]   Lewis H., Technological Risk: What are the real dangers, if any, of toxic chemicals, the greenhouse effect, microwave radiation, nuclear power, air travel, automobile travel, carcinogens of all kinds, and other threats to our peace of mind?, 1990

[81]   Libet B. et al., Readiness potentials preceding unrestricted spontaneous pre-planned voluntary acts, 1983

[82]   Lippert A. et al,: Vom Leben und Sterben des Ötztaler Gletschermannes: Neue medizinische und archäologische Erkenntnisse, 2007

[83]   McCammon I., Heuristic Traps in Recreational Avalanche Accidents: Evidence and Implications, 2004

[84]   McCarthy W., A Brief Look at Climate Change and Global Warming, 2011

[85]   Meehl P., Clinical Versus Statistical Prediction: A Theoretical Analysis and a Review of the Evidence, 1954

[86]   Milgram S., Obedience to Authority: An Experimental View, 1974

[87]   Miller G., The Magical Number Seven, Plus or Minus Two: Some Limits on our Capacity for Processing Information, 1956

[88]   Ministry of Road Transport and Highways Transport Research Wing, Road Accidents in India 2013

[89]   Morgan, M., Fischhoff B. et al., Risk Communication: A Mental Models Approach, 2001

[90]   Mumford J., Road deaths in developing countries; the challenge of dysfunctional roads, 2008

[91]   Munter W., 3x3 Lawinen: Risikomanagement im Wintersport, 2003

[92]   NASA/SP-2011-3421, Probabilistic Risk Assessment Procedures Guide for NASA Managers and Practitioners

[93]   NUREG/BR-0359, Modeling Potential Reactor Accident Consequences, USNRC, 2012

[94]   NUREG/CR-1250, Three Mile Island; A Report to the Commissioners and to the Public, 1980

[95]   Nyhan B., Reifler J, When Corrections Fail: The Persistence of Political Misperceptions, 2010

[96]   Peltzman S., The Effects of Automobile Safety Regulation, 1975

[97]   Perrow C., Normal Accidents: Living with High-Risk Technologies, 1984

[98]   Peter L., Hull R., The Peter Principle: Why Things Always Go Wrong, 1969

[99]   Petrovski H., To Engineer Is Human: The Role of Failure in Successful Design, 1985

[100]  PIARC 05.04.B, Road Safety in Tunnels, 1995

[101]  PIARC 05.16.B, Systems and Equipment for Fire and Smoke Control in Road Tunnels, 2006

[102]  PIARC 2010R1, Towards Development of a Risk Management Approach

[103]  PIARC 2012R23EN, Current Practice for Risk Evaluation for Road Tunnels

[104] PIARC 2012R30EN, Social Acceptance of Risks and their Perception

[105] Pinker S., The better Angels of our Nature: Why Violence has declined, 2011

[106] Popper K., Logik der Forschung, 1935 (The Logic of Scientific Discovery, 1959)

[107] Popper K., The Open Society and Its Enemies, 1945

[108] Pospisil P., Reducing Costs and Improving Safety of Road Tunnels, 2011

[109] Pospisil P., Road Tunnel Ventilation: Compendium and Practical Guideline, 2013

[110] Price T., What Went Wrong: Oil Refinery Disaster, Popular Mechanics, 2005

[111] Putt A., Putt's Law and the Successful Technocrat: How to Win in the Information Age, 2006

[112] Quarantelli E., Disasters: Theory and Research, 1978

[113] Reason J., Human Error, 1990

[114] Reason J., The Human Contribution: Unsafe Acts, Accidents and Heroic Recovers, 2008

[115] Renn O., Risk Governance: Coping with Uncertainties in a Complex World, 2008

[116] Report of Japanese Government to IAEA Ministerial Conference on Nuclear Safety - Accident at TEPCO's Fukushima Nuclear Power Stations, 2011 (www.iaea.org/newscenter/focus/fukushima/japan-report/)

[117] Report of the Presidential Commission on the Space Shuttle Challenger Accident,1986 (history.nasa.gov/rogersrep/genindex.htm)

[118] Ripley A., The Unthinkable: Who Survives When Disaster Strikes - and Why, 2008

[119] Roser M., Ethnographic and Archaeological Evidence on Violent Deaths', 2016 (OurWorldInData.org)

[120] Schwartz G., The Myth of the Ford Pinto Case, 1991

[121] Sherif M., A study of some social factors in perception, 1935

[122] Slovic P. et al., The Feeling of Risk: New Perspectives on Risk Perception, 2010

[123] Slovic P., Perception of Risk, 1987

[124] Spitzer M., Digitale Demenz: Wie wir uns und unsere Kinder um den Verstand bringen (Digital Dementia), 2012

[125] Stranks J., Human Factors and Behavioural Safety, 2007

[126] Sunstein C., Risk and Reason: Safety, Law and the Environment, 2002

[127] Taleb N., Douady R., Mathematical Definition, Mapping, and Detection of (Anti)Fragility, 2013

[128] Taleb N., Fooled by Randomness: The Hidden Role of Chance in Life and in the Markets, 2001

[129] Taleb N., The Black Swan: The Impact of the Highly Improbable, 2007

[130] Tavris C., Aronson E., Mistakes Were Made (But not by Me): Why We Justify Foolish Beliefs, Bad Decisions, and Hurtful Acts, 2007

[131] Tengs T. et al., Five-Hundred Life-Saving Interventions and Their Cost-effectiveness, 1995

[132] Tetlock P., Expert Political Judgement: How Good Is It? How Can We Know? 2005

[133] Three Mile Island Accident, Backgrounder, USNRC (http://www.nrc.gov/reading-rm/doc-collections/fact-sheets/3mile-isle.html)

[134] Tomasetti C., Vogelstein B., Variation in cancer risk among tissues can be explained by the number of stem cell divisions, 2015

[135] Tversky A., Kahnemann D., The Framing of Decisions and the Psychology of Choice, 1981

[136] Viscusi K., Gayer T., Safety at Any Price? 2002

[137] Wahlberg A., Risk Perception and the Media, 2000

[138] Wilde G., Target Risk 3, 2014

[139] Williams M. et al., Environmental Impact of the 73 ka Toba super-eruption in South Asia, 2009

[140] Wisner B. et al., At Risk: Natural hazards, people's vulnerability and disasters, 1994

[141] World Nuclear Association, Chernobyl Accident 1986 (www.world-nuclear.org/info/Safety-and-Security/Safety-of-Plants/Chernobyl-Accident/)

[142] Zadeh L., Kacprzyk J., Fuzzy Logic for the Management of Uncertainty, 1992

[143] Zimbardo P. et al., Stanford Prison Experiment 1971 (http://purl.stanford.edu/vx097ry2810)

[144] Zweifel B., Who Is Involved in Avalanche Accidents, 2012

## 14.2  Internet Links

Search on the Internet provides plenty of information. Internet links are dynamic and constantly changing. Evaluating their trustworthiness, and application to specific issues is strenuous. In the age of overabundance of information, choosing and evaluating the relevant facts is a difficult task.

Nevertheless, a short excerpt of sources for statistical data about probabilities of individual risks and mortality rates is provided:

*www.who.int/mediacentre/factsheets/fs310/en/*

*www.who.int/gho/en/*

*https://www.census.gov/compendia/statab/cats/
births_deaths_marriages_divorces.html*

*www.cdc.gov/nchs/deaths.htm*

*www.nsc.org*

*www.statistics.gov.uk/hub/health-social-care/
health-of-the-population/causes-of-death*

*www.abs.gov.au/ausstats/abs@.nsf/mf/3303.0/*

*https://www.destatis.de/de/zahlenfakten/gesellschaftstaat/
gesundheit/todesursachen/todesursachen.html*

*www.bfs.admin.ch/bfs/portal/de/index/themen/14/02/04/
key/01.html*

*http://www.cancer.org/research/cancerfactsstatistics/*

*http://www.piarc.org/en/knowledge-base/road-safety/*

*http://www.tunnels.piarc.org/en/safety/experience.htm.*

*https://ourworldindata.org/*

## 14.3   Author's Notes and Acknowledgments

*'Everything doesn't have to be written, and not all that is written may be under-stood'.*

This book was not written in an academic environment and financed by the taxpayer, but parallel to ongoing professional work as consultant, designer and site supervisor. The text was written in English for an international audience, although English is neither my mother tongue nor my common language. I beg the reader to forgive my simple, technically oriented writing with limited vo-cabulary and without poetic elegance. The sources for the thoughts and state-ments in this book are my professional engineering background, my experi-ence as adventurer and outdoorsman, exchange of views with colleagues and fellow professionals, and many books, articles and other publications, written by wiser women and men, for instance those listed in the bibliography (chapter 14.1). While carrying out an extensive literature research I found out that many of my thoughts had already been written down by other authors. Rather than re-write or plagiarize some very good books, I tried to communicate my own thoughts to the general public, applying simple logic beyond the usual bureau-cratic risk management approach, based on my own, personal experience and professional insight. Many of the statements in this book have been previously published in some of my presentations, articles, papers and technical reports for a professional audience[38].

Life is unpredictable, and risk is an important part of life. Like most men, I took many risks at a young age, often going to the limit and sometimes beyond, learning from success and even more from faults, slips and accidents. I have worked with brilliant theoretical academics, as well as with experienced, hands-on practitioners, and discussed the issues with friends who worked in other fields of activity. Working under a constant pressure of keeping to deadlines and financial restrictions, I observed an increasing discrepancy between offi-cially communicated quality and safety issues, and practical application. Many questions aroused: Why do things go wrong? How can I avoid mistakes in my own work and in that of others involved? Why do we stick to bad decisions even after admitting that they were erroneous? Why are funds wasted for use-less technical measures, but the involved professionals have to work for low wages under cost and time pressure?

---

[38] http://www.p-i.ch/cms/pages/english/publications.php

Many risks arise because of lack of communication and understanding between theoreticians and practitioners. I have worked on both sides, and experienced at first hand the difficulties of understanding between different viewpoints, but that helped me to broaden my horizon.

This book would not have been written without my father, a real critical spirit. After the invasion of Czechoslovakia by the Soviet Union and their allies, he decided to emigrate with his family to Switzerland, giving me the opportunity of growing up in a democracy with the right of free speech. Thus, Communist dictatorship had always been my concept of the enemy. Stupid people without education, but a strong urge for power, were in charge of that totalitarian regime. Statements that were obviously absurd were propagated as official ideology. The whole system was built on a basis of lies, but almost all people participated in one way or another, even when complaining in secret. Since then, times have changed. Communist dictatorships have disappeared, with a few exceptions. What really gives rise to concern is that I observe the attributes of totalitarianism in public authorities and big companies in Western democracies. Such are for instance excessive bureaucracy, the demand to strictly follow orders without reflection, and executives with little professional competence, but almost unlimited power, deciding about the investment of huge sums of taxpayer's or shareholder's money in high stake risk issues, without really taking responsibility. Therefore, this book shall be a contribution to the struggle against the ever-present danger of relapse into the medieval Dark Age.

I want to thank my family, friends and colleagues for mental support and patience, and the honorable employers and clients that enabled me to do interesting work, from which I derived my professional experience. Without my daughter Carolina, this would have remained a plain technical paper for engineers, and not a book that is (hopefully) understandable to the general public. Special thanks to André Müller, an English teacher, for professional lecture.

If you find any obvious faults or contradictory information in this book, let me know by mail to books@p-i.ch.